SHAPING SMART MOBILITY FUTURES

SHAPING SMART MOBILITY FUTURES: GOVERNANCE AND POLICY INSTRUMENTS IN TIMES OF SUSTAINABILITY TRANSITIONS

EDITED BY

ALEXANDER PAULSSON

Lund University, Sweden

and

CLAUS HEDEGAARD SØRENSEN

Swedish National Road and Transport Research Institute (VTI), Sweden

emerald
PUBLISHING

United Kingdom – North America – Japan – India – Malaysia – China

Emerald Publishing Limited
Howard House, Wagon Lane, Bingley BD16 1WA, UK

First edition 2020

British Library Cataloguing in Publication Data
A catalogue record for this book is available from the British Library

ISBN: 978-1-83982-651-1 (Print)
ISBN: 978-1-83982-650-4 (Online)
ISBN: 978-1-83982-652-8 (Epub)

Printed and bound by CPI Group (UK) Ltd, Croydon, CR0 4YY

ISOQAR
REGISTERED
Certificate Number 1985
ISO 14001

ISOQAR certified Management System, awarded to Emerald for adherence to Environmental standard ISO 14001:2004.

INVESTOR IN PEOPLE

Contents

List of Figures and Tables

Chapter 9

Chapter 10

List of Contributors

David Ashmore	Faculty of Architecture, Building & Planning, University of Melbourne, Australia
Carey Curtis	Curtin University, Australia
Iain Docherty	Institute for Advanced Studies, University of Stirling, UK
Albin Engholm	Integrated Transport Research Lab, KTH Royal Institute of Technology, Sweden
Nils Fearnley	Institute of Transport Economics, Norway
John Hultén	The Swedish Knowledge Centre for Public Transport, (K2) Sweden
Johanna Jussila Hammes	Swedish National Road and Transport Research Institute (VTI), Sweden
Jamil Khan	Lund University, Environmental and Energy Systems Studies, Sweden
Ida Kristoffersson	Swedish National Road and Transport Research Institute (VTI), Sweden
Annica Kronsell	School of Global Studies, University of Gothenburg, Sweden
Crystal Legacy	Faculty of Architecture, Building & Planning, University of Melbourne, Australia
Sam McLeod	Curtin University, Australia
Ioanna Moscholidou	Institute of Transport Studies, University of Leeds, UK
Dalia Mukhtar-Landgren	Department of Political Science, Lund University, Sweden
Alexander Paulsson	The Swedish Knowledge Centre for Public Transport, (K2) Sweden and Lund University School of Economics and Management, Sweden

Anna Pernestål	Integrated Transport Research Lab, KTH Royal Institute of Technology, Sweden
Fredrik Pettersson	Lund University, Transport & Roads, Sweden
Louise Reardon	INLOGOV, University of Birmingham, UK
Claus Hedegaard Sørensen	The Swedish Knowledge Centre for Public Transport (K2), Sweden and Swedish National Road and Transport Research Institute (VTI), Sweden
John Stone	Faculty of Architecture, Building & Planning, University of Melbourne, Australia
Anna Wallsten	Swedish National Road and Transport Research Institute (VTI), Sweden

About the Contributors

David Ashmore is a Researcher at the University of Melbourne. He recently completed his doctorate, which examines the symbolic aspects of transport choice across different cultures. His professional background is in transport regulation and procurement; he has worked for consulting firms, universities, and the civil service.

Carey Curtis is a Professor of City Planning and Transport at Curtin University, a Director of Urbanet Research Network, and a Guest Professor at the K2/University of Lund. Her research interests include city form and structure, transit-oriented development, personal travel behaviour, accessibility planning, institutional barriers to sustainable transport, governance, and transport policy.

Iain Docherty is the Dean of the Institute of Advanced Studies and a Professor of Public Policy and Governance at the University of Stirling, Scotland, UK. His research and teaching address the interconnecting issues of public administration, institutional change, and city and regional competitiveness, with particular emphasis on the structures and processes of local and regional governance, policies for delivering improved economic performance and environmental sustainability, and the development and implementation of strategic planning and transport policies. His recent books include *The Transport Debate* and *Transport Matters*, both with long-term research collaborator Professor Jon Shaw.

Albin Engholm is a PhD candidate at Integrated Transport Research Lab at KTH Royal Institute of Technology in Stockholm, Sweden. His research interests include the long-term impacts of driverless vehicles on the transport system and society by combining future studies methods with various modelling approaches. He is particularly interested in how driving automation creates opportunities and challenges for making freight transport systems more sustainable.

Nils Fearnley is a Transport Economist and a Senior Researcher in urban and passenger transport. He is currently a Chief Researcher for the Group Market and Governance at the Institute of Transport Economics. His research interests include governance, regulation, and financing of passenger transport; economic and social aspects of transport; market analysis; transport appraisal; transport policy; and transport statistics.

John Hultén is Director of K2 the Swedish Knowledge Centre for Public Transport. He has more than 15 years of experience from working with transportation and mobility in Sweden, for example, within the Swedish Road Administration, the Swedish Transport Administration, and the Ministry of Transportation. He holds a PhD in Political Science from Lund University and has conducted research on the politics of transportation planning, governance, and funding.

Johanna Jussila Hammes, has a PhD in Environmental Economics from the University of Gothenburg. She has been working at VTI since 2009, publishing analyses of the political economy of infrastructure investment planning, and policy instruments for biofuels, and infrastructure investment. She is currently working with local-level policies for sustainable transport in Sweden, the behaviour of civil servants in the decision-making process, electrification of roads for heavy transport, and the needs for changing present policies to facilitate a transition from the present, and fossil-based transport system towards a sustainable one.

Jamil Khan is an Associate Professor at Environmental and Energy Systems Studies at Lund University. He has researched on climate politics, low carbon transitions, and sustainable transport planning for 20 years. He has published widely in the field and has contributed to books such as *Rethinking the Green State* (2015) and *Sustainability and the Political Economy of Welfare* (2016).

Ida Kristoffersson has a PhD in Transport Science from KTH Royal Institute of Technology. She has been working as a Researcher at VTI since 2016 and in 2019 became a Senior Research Leader in the field modelling and analysis of passenger transport. Her main research area is in development of travel demand models to meet the needs for evaluation of new policies and innovations given the increased attention to sustainability and digitalisation of the transport sector. She has published book chapters and journal articles in fields such as modelling and effects of *congestion* charges, as well as future scenarios and long-term effects of self-driving vehicles.

Annica Kronsell is Professor of Political Science and Chair of Environmental Social Science at the School of Global Studies, Gothenburg University. She is interested in how public institutions can govern climate and sustainability issues. As part of various multidisciplinary consortia, she has studied different dimensions of climate governance in the Scandinavian context and published articles and books on the green public sector and environmental governance and on municipalities in experimental governance and they include: *Rethinking the Green State. Environmental Governance towards Environmental and Sustainability Transitions* (with Bäckstrand, Routledge, 2015) and 'The Green Decarbonised State and Industrial Governance' with Hildingsson and Khan in *Environmental Politics*. She also uses feminist theorising to study power relations in transport governance with publications such as 'Investigating the Link Between Transport Sustainability and the Representation of Women in Swedish Local Committees' (2019) with Winslott Hiselius, Dymén, and Smidfelt in *Sustainability*.

Crystal Legacy is a Senior Lecturer in Urban Planning at the University of Melbourne. Her research examines questions of urban conflict and citizen engagement with a current focus on the role of the citizen in contested transport processes in Australian and Canadian cities. She is the Co-editor of *Instruments of Planning: Tensions and Challenge for more Equitable and Sustainable Cities* (Routledge, 2016).

Sam McLeod is a Researcher at Curtin University, and a practicing Transport Planner at consultancy firm GHD. He has contributed to a range of academic and applied research, with particular focus on strategic metropolitan planning, transport planning and economics, planning for freight, managing uncertainty and change, and evidence-based planning. He holds qualifications in Urban and Regional Planning and Project Management.

Ioanna Moscholidou is a PhD student at the Institute for Transport Studies, University of Leeds. She is researching how cities govern smart mobility services and how public authorities can steer services towards achieving local sustainable transport goals.

Dalia Mukhtar-Landgren is a Senior Lecturer in Political Science at Lund University, Sweden. Her research interests are centred around relations of power and politics in urban planning and development, as well as in local development politics at large. Her recent publications include studies of public sector projectification, experimentation, and local innovation work. She is currently engaged in research projects on testbed planning, urban experimentation, smart mobility, and processes of local innovation and development practices.

Alexander Paulsson is a Lecturer at Lund University School of Economics and Management and a Researcher at the Swedish Knowledge Centre for Public Transport. He is currently doing research on the governance of new forms of mobility and the marketisation of public transport as well as political economies of post-growth societies. His research interests are broadly within the areas of organisation studies, science, and technology studies as well as ecological economics. He has recently edited (with S. Barca and E. Chertkovskaya) *Towards a Political Economy of Degrowth* (Rowman and Littlefield Publishers, 2019).

Anna Pernestål received her PhD in Systems Engineering in 2009. She has been active in transportation industry for more than 15 years, and has had managing positions within both road and rail sectors. Currently, she is the Director for the research centre Integrated Transport Research Lab, where the focus is building knowledge about how new technology such as digitalisation and automation can contribute to a sustainable transportation system. Her research interests are within system level analysis and design of the transportation system.

Fredrik Pettersson is Associate Senior Lecturer at Transport and Roads, Lund University, and also involved in K2 the Swedish Knowledge Centre for Public Transport. His research interest is in the dynamics between different levels of decision-making and different organisations in the transition to a more sustainable transport system. In the last decade, he has published research on national level transport policy-making as well as planning and decision-making processes at local and regional levels.

Louise Reardon is a Lecturer in Governance and Public Policy at the Institute of Local Government Studies, University of Birmingham. Her research is at the forefront of knowledge at the interdisciplinary nexus of governance and public policy, transport, and wellbeing research. In particular, her research focusses on the role multi-level governance and institutional networks play in influencing policy agendas and shaping policy outcomes. She recently co-edited the book *Governance of the Smart Mobility Transition* (Emerald, 2018). The book presents an agenda for future research and policy action around the role and impact of governance in relation to smart mobility. She is the Co-editor of the journal *Local Government Studies* and Co-chair of the Governance and Decision Making Processes Special Interest Group of the World Conference on Transport Research Society.

Claus Hedegaard Sørensen is a Research Leader at the Swedish Knowledge Centre for Public Transport (K2) and a Senior Researcher at Swedish National Road and Transport Research Institute (VTI). He is conducting research on transport governance, and his research has mainly focussed on environmental policy integration in transport; national transport planning; organisation and collaboration within public transport; as well as the use and role of knowledge in transport policy-making. The last couple of years he has mostly researched and published on the governance of smart mobility.

John Stone is a Senior Lecturer in Transport Planning at the University of Melbourne. His research explores the political and institutional context for variation in international transport planning practice, with a focus on cities in Australia, Canada, and German-speaking Europe. He has also worked in local government and as a community advocate for sustainable transport.

Anna Wallsten is a Postdoctoral Researcher at the Swedish National Road and Transport Research Institute (VTI). With an interdisciplinary background, she addresses research within the field of sustainable transitions; citizen engagements; science and technology studies; and future studies. She holds a Doctoral degree in Technology and Social Change. Her previous work concerns visions of smart grids, and the tensions that occur when such prospects are translated into practice within demonstration projects. Her current research focusses on issues concerning the digitalisation of the transport system, emerging digitally supported transport solutions, and the institutional capacity of public actors to steer the development towards achieving long-term societal objectives.

Preface

By the beginning of 2018, we were leading a number of research projects affiliated to the Swedish Knowledge Centre for Public Transport (K2). All the projects in one way or another focussed on the governance of smart mobility. Being engaged in these projects, we saw a need for contemplating these issues in another way than allowed by academic journal papers or presentations at conferences, seminars, and workshops. Editing an anthology like this book was an attractive opportunity, as it provided a possibility for showing the complexity of the issue and the diversity of perspectives.

The result is this book. During the spring of 2019, we invited potential chapter authors to a seminar at K2 in Lund, Sweden to be held in September 2019. This seminar provided a valuable opportunity to discuss ideas, and drafts of all chapters were critically reviewed. Some of the chapters were authored by researchers involved directly in the above-mentioned research projects. Other chapters were penned by members of scientific advisory groups connected to these projects. And a third group of chapters were written by other colleagues involved in similar research projects.

We would like to express our gratitude to all chapter authors for their engagement in this project, as well as to colleagues at the K2 Centre, as this has formed an inspiring environment for the work. The research projects making this book possible were funded by the Swedish Energy Agency, The Swedish Innovation Agency (Vinnova), and The Swedish Knowledge Centre for Public Transport (K2).

Last but not least, we would like to thank our families for support and understanding throughout this process.

Lund, Sweden, January 2020
Alexander Paulsson and Claus Hedegaard Sørensen

Chapter 1

Smart Mobility and Policy Instruments: Broadened Definitions and Critical Understandings

Alexander Paulsson and Claus Hedegaard Sørensen

ABSTRACT

The point of departure of this book is that smart mobility will only be developed in a desired direction and fulfil societal objectives if it is steered in that direction. The market, left to itself, will most certainly not deliver on these objectives. This message has been conveyed extensively in recent literature, but this book aims to take this discussion one step further by focussing on what governance of smart mobility looks like today and in the future. In this introductory chapter, the authors provide a framework of different understandings of policy instruments, how they are selected, developed and used. After the array of policy instruments within the transport sector has been extensively discussed, the authors turn to discussing a broader understanding of policy instruments found within political science and political sociology. In doing so, this book contributes to the critical scholarship on policy instruments, while exploring the why, the how and the what of policy instruments in relation to smart mobility. The chapter closes with a brief introduction to the structure of the book as well as a description of the content of each chapter.

Keywords: Smart mobility; policy instruments; governance; transport policy; political objectives; sustainable mobility

Introduction

In recent years, the advent of autonomous vehicles, the roll-out of electrification and the introduction of shared mobility solutions have influenced the

Shaping Smart Mobility Futures:
Governance and Policy Instruments in times of Sustainability Transitions, 1–16

doi:10.1108/978-1-83982-650-420201001

public debate as well as academic conferences and publications within the field of mobility and transport. While these developments differ in terms of both technologies and business models, the notions of 'smart mobility' and 'smart transport' are increasingly applied as synonyms of mobility and transport futures. Included in smart mobility futures are, for example, autonomous and connected vehicles, shared platform-based mobility such as car-sharing, bike-sharing, ride-sharing, combined mobility like Mobility as a Service, as well as electrification (Papa & Lauwers, 2015). The term 'smart' is currently used in connection with a multitude of devices (smart phone, smart television and smart card), and '"smart" is the order of the day' as one author stressed (Lyons, 2018, p. 2). There is considerable political and industry-related enthusiasm for this development (Fagnant & Kockelman, 2015; Herrmann, Brenner, & Stadler, 2018; Seba, 2014). However, public debate and academic publications to an increasing extent stress that smart mobility might also be counterproductive, for example, regarding sustainability, congestion, attractiveness of cities, etc. (Docherty, Marsden, & Anable, 2018; Marsden & Reardon, 2018; Schiller, 2016). It is indeed contestable whether 'smart mobility' really *is* smart.

Landmark studies within the field include several transport model studies focussing on specific cities carried out by International Transport Forum. Studies have been conducted for Lisbon, Auckland, Helsinki, and Dublin (International Transport Forum, 2017a, 2017b, 2017c, 2018) and have inspired similar studies in other cities (COWI & PTV, 2019). The main conclusion from these studies is that sharing is necessary to achieve societal objectives using new technologies, and sharing in combination with public transport can contribute to reducing the number of cars, traffic volume, parking spaces, congestion, pollution, CO_2 emissions, etc. Though the studies have been criticised for applying unrealistic assumptions (Docherty et al., 2018), the studies have both inspired and prompted public authorities (COWI & PTV, 2019) and corporations (e.g. Pietzsch, 2018) to launch initiatives on shared and combined mobility.

A Need for Governance?

Based in the modelling studies mentioned above as well as several real-world pilots and tests, an emerging body of literature has stressed the need for governance of 'smart' mobility, not least during the transition process (Docherty et al., 2018; Finger & Audouin, 2018; Marsden & Reardon, 2018). Docherty et al. (2018) have argued that smart mobility can only be developed in a desired direction and fulfil societal objectives if it is steered in that direction. The market, left to itself, will most certainly not deliver on these objectives. Finger and Audouin (2018) as well as Lyons (2018) have stressed a need to align smart mobility with the sustainability paradigm. Docherty et al. (2018) have also argued that there is, in the transport sector, a tradition for public sector involvement that should be applied to govern the transition, while Reardon and Marsden (2018) express an urge to use the current window of opportunity for deliberate considerations and debates on societal goals, suggesting a need for cautiousness before entering into possible, disruptive societal changes.

This book is written and published in a time of climate crisis that probably represents one of the biggest global challenges in the history of mankind. A huge gap exists between, on the one hand, how scientists within the climate field express the need for policy initiatives at all levels of society (Lenton et al., 2019), and on the other hand, the actually implemented policy initiatives, which so far seem insufficient. When it comes to transport and mobility, smart mobility is often envisaged as a solution that enables highly mobile societies with a limited carbon footprint, because mobility is expected to be electrified, shared and more efficient (Seba, 2014). For that reason, achieving smart mobility is often expressed as a goal on its own, and some of the literature on smart mobility is focussed on how to introduce and implement smart mobility solutions, thus overcoming legal obstacles and popular resistance (Bjelfvenstam, 2018; Herrmann et al., 2018). However, the chapters in this volume do not subscribe a priori to this optimistic approach to smart mobility, since smart mobility might also have undesired and unintended consequences. When governance and policy instruments are analysed in this volume, the ambition is not to discuss how to implement smart mobility in the most efficient way, but rather to discuss how smart mobility can be governed so that broader sustainability goals or other societal goals are achieved.

The demand for governance and policy instruments in relation to the smart mobility transition is strongly linked to societal goals introduced in recent years at various levels of society. The 17 Sustainable Development Goals are one example (Hildebrandt, 2016; United Nations, 2019) highlighting global ambitions to combat poverty and climate change and to establish wealthy and more equal societies. The Sustainable Development Goals are inspiring national and local public authorities as well as corporations, which in some cases even adopt the SDGs in their vision and mission statements (Ali, Hussain, Zhang, Nurunnabi, & Li, 2018). At the national level, transport policy objectives are established in many countries, including ambitions regarding climate, safety, accessibility, environment and growth (Sørensen & Gudmundsson, 2010). Within other sectors, overarching goals are formulated which also aim to impact decisions within the field of transport. At the regional and municipal levels, a similar trend is observed, a remarkable example being the cities gathered in networks like C40 and ICLEI to achieve climate and sustainability goals (C40, 2019; ICLEI, 2019).

What is a Policy Instrument?

Despite the assumption that political involvement is necessary, surprisingly little is said in the literature about what such governance would look like. Policy instruments are one way that governance is carried out or achieved, and as such, are often imbued with a means-end rationality. While policy instruments are often designed to achieve specific objectives, policy instruments may also be developed and used for fulfiling a wide range of other objectives. The characteristics of the objective generally influence how the policy instrument is designed and implemented (Edmondson, Kern, & Rogge, 2018; Rogge & Reichardt, 2016). Also, the context of a specific policy instrument matters (Sørensen & Paulsson, 2019, in press). For example, during times of relative stability, policy instruments may build on previously implemented instruments, become institutionalised and even mainstream, as it were. During times of rapid transition or unexpected and disruptive socio-technological change, newly designed policy instruments may become obsolete and older instruments may turn out to be counter-productive. In short, both the speed and the direction of socio-technological changes may impact which policy instruments are regarded as suitable and relevant for achieving the desired objectives (see, e.g., Stead & Vaddadi, 2019). In fact, the choice of instrument generally reflects the political and administrative elites' ideas about the relationship between the governing and the governed (Gössling & Cohen, 2014). Since the choice and design of policy instruments are shaped by political and administrative elites' ideas as well as broader socio-political developments, policy instruments are often highly politicised (Howlett, 2009; Howlett & Ramesh, 1993; Stead, 2018).

The ambition of this book is to take the call for governance of smart mobility one step further by considering the policy instruments used today and the instruments that might be used in shaping smart mobility futures, thus enabling them to meet the societal objectives like tackling the ongoing climate crisis and achieve the goal of sustainable mobility. As of today, policy instruments used in the transport sector include, but are not limited to, a range of taxes and fees, as well as legislation on traffic and vehicles that includes parking restrictions and land-use planning. In futures of smart mobility, some of these policy instruments might be weakened as the technology makes them either obsolete or redundant. For example, autonomous cars presumably would not have to care about parking restrictions, because they can continue circulating or park outside the city centre. With changing ownership structures and shared forms of mobility solutions, some taxes on vehicle ownership and fuel consumption may prove to be weak instruments to shape travel behaviour. Yet, other policy instruments might be strengthened as the socio-technological changes are more aligned with the objectives of the instruments, for example, taxes or fees based on kilometres travelled. There might also be completely new forms of instruments emerging from the socio-technological changes that public authorities may try to use, or they might want to recalibrate current policy instruments due to the emergence of smart mobility. We see this focus on policy instruments as a new topic within the literature on smart mobility, since it goes beyond the plain call for governance to secure public values.

A framework is developed below to gain an overview of the policy instruments presently used in the transport sector. However, we want to stress that in this book, there is no normative standpoint as regards certain types of policy instruments, nor do we endorse a certain type of knowledge or framework about policy instruments. Instead, the point of departure is broad, and we suggest that policy instruments should be understood as 'techniques of governance that, one way or another, involve the utilization of state authority or its conscious limitation' (Howlett, 2005, p. 31). Rather than resorting to functionalist explanations of policy instruments, that is, as rational tools used by governments to achieve clearly defined objectives, we include a range of conceptualisations, from economics and political science to sociology. In doing so, we contribute to the literature by problematising the definition of policy instruments and how they develop, operate and impact their target audience.

In light of such a broad understanding and the multitude of conceptualisations included in this book, we will first discuss policy instruments and instrument categorisations applied within the field of transport, and we will then turn to more recent research, advocating a broader understanding of policy instruments.

Policy Instruments Used to Govern Transport

In the transport sector, the discussion on policy instruments remains vivid and the debated socio-technological changes associated with smart mobility, as well as the policy ambitions to reach the climate goals, has brought to light the need to consider the importance and effectiveness of the policy instruments used in the sector. Therefore, we will now discuss the instruments in use. The purpose of these instruments is often to steer the development towards societal objectives, for example, reducing travel demand, shifting travel from car to walking, cycling and public transport, as well as developing or using more energy-efficient fuels.

As discussed above, the purpose of using policy instruments is to influence the decisions and behaviours of a subsector of society or a predefined target group in order to achieve certain intended effects. In the transport sector alone, there is a plethora of instruments. According to one estimate, there are up to 60 different types of instruments in this sector (Institute for Transport Studies, 2009). Because of this multitude and lack of overview, there have been several attempts to categorise the policy instruments used in the transport sector. Santos, Behrendt, and Teytelboym (2010) divide instruments for sustainable road transport into three categories: physical instruments, soft instruments and knowledge as an instrument. Physical instruments include policies that affect the built environment and infrastructure, for example, raising land and capacity for the construction of roads. Soft tools and knowledge aim to change behavioural patterns through targeted information and marketing, as well as norms and standards.

The distinction between hard and physical instruments on the one hand and soft and 'non-physical' instruments on the other is relatively established in the transport sector, but the analytical value of the distinction can be discussed. Policies that *directly* intend to steer behavioural patterns in a desired direction, such as information campaigns or marketing, are obviously non-physical. But measures aiming to change behaviour often also involve changes in the physical environment. Pedestrian crossings, speed bumps or other measures in the physical infrastructure are some examples of this. Transforming and changing the built environment and thereby increasing accessibility is a 'physical' policy instrument. At the same time, such instruments operate in a context where administrative or non-physical measures are inter-dependent. Pedestrianisation and car-free zones are two such policies often used by cities promoting sustainable mobility. Changing the use of road space so that buses and tramways are prioritised over cars are similar examples of cities promoting sustainable forms of mobility (Petterson-Löfstedt & Sørensen, 2019, in press). Such changes are primarily based on administrative decisions but also include changes both in the production of urban space and more specifically in the use of existing road space. In short, administrative policy instruments and interventions in the physical and built environment often interact with each other.

In the transport sector, it is also common to distinguish between financial, administrative and informative instruments (Dickinson & Wretstrand, 2015). We will therefore briefly discuss and contextualise these categories below.

Financial instruments are based on financial incentives as well as monetary costs and benefits. Behaviour is affected to the extent that actors are motivated by or place monetary value on their behaviours and decisions. Because actors attribute different values to costs and benefits, for example, depending on relative budgetary constraints and priorities, the effects of financial policy instruments are highly differentiated. Fees, taxes and subsidies are prominent and common instruments (Frey, 2003). Fees, taxes and subsidies have the potential to manage behaviour as well as to finance and internalise costs associated with the use of the transport system (Nash & Matthews, 2013). However, the introduction of fees, taxes and subsidies affect different groups in different ways. For households with relatively high incomes, the effects may be marginal, while the effects for households with lower incomes will be comparatively substantial. Governments also use various forms of subsidies in the transport sector to promote certain desired behaviours and decisions (Notteboom, 2013).

Administrative instruments are instruments based on the ability of public actors to impose restrictions, requirements, certifications, permits or formal decisions. Administrative instruments include everything from legislation and regulations to policies and recommendations. Examples of administrative instruments in the road infrastructure network are speed restrictions and parking restrictions. Failure to comply with legislation and regulations can result in financial penalties, such as fines, as well as administrative penalties, such as suspension of permits and prohibitions. Failure to comply with policy usually causes suspension of permits or certifications (e.g. Dowling, 2018; Rodrigue, 2013).

Informative instruments refer to instruments intended to influence behaviour and traffic flows through knowledge, communication and nudging. *Mobility management* is an example of a policy instrument that incorporates informational elements. In practice, mobility management may be about informing and planning for a more efficient private car use, to encourage car users to share rides or vehicles, or use public transport or cycle and walk. Mobility management is the instrument that requires the smallest interventions to potentially influence the mobility practices and behaviours and so optimise the capacity of the transport system. However, it is an instrument that may lead to controversies, not least because it can be perceived to put individual liberty and freedom of movement into question (Hrelja, Isaksson, & Richardson, 2013). Intelligent transport systems (ITS), which are already in use in several places, are expected to grow in usage because of the roll-out of automated and connected vehicles (see Hopkins & Schwanen, 2018; Janecki, 2011).

In addition to these three categories of instruments, there is a fourth policy instrument: *research and development*. Research, development and demonstration projects are ways of 'correcting' market failures, since the market produces too little knowledge and innovation when left to its own devices. Knowledge about the effects of new technology and innovations is often seen as a prerequisite for being able to achieve different environmental goals in the long term. This justifies the use of public funds to steer knowledge development in a desirable direction through pilot experiments, test beds or earmarked research funding.

Governments at various levels – be it national, federal, regional or local – are all involved in governing the transport sector using different policy instruments. Cities are central to the transport system as they accommodate loads of cars and freight trucks, but cities are also innovative when it comes to policies curbing car use. With the help of land-use policy and traffic strategies, cities are trying to steer urban mobility in a sustainable direction, not least by combining 'physical' instruments with changes in the built environment. According to a study by Grazi and van den Bergh (2008), land-use policy is a superior policy instrument, as it sets out the material framework and the possible effects of the other policy instruments: whether they are financial, administrative and/or informational policy instruments.

The number of policy instruments in the transport sector is impressive, spanning from traffic rules via huge infrastructure investments to land use planning, and the broader scholarly debate on governance and policy instruments is similarly rich and encompasses an extensive literature. Although we do not intend to contribute to the discussion on the categorisation of policy instruments, we believe that we can add perspectives to the discussion by trying to broaden the understanding of policy instruments and by highlighting the limits of such instruments.

Towards a Broader Understanding

The research on policy instruments can be divided into two camps. On the one hand, there is a distinct normative approach that tries to describe how the state, by adopting 'the right' policy instruments, could plan, control or steer the development of society or specific subsectors. On the other hand, there is a more descriptive approach that attempts to map the instruments or mechanisms by which the state de facto governs society or aspects of it. In the middle of these lies the pragmatic approach, which aims to describe how the state can steer subsectors of society in certain directions, but doing so based on the policy instruments, tools and mechanisms that are already in use and where the state already possesses the required knowledge and competence. With the book *The Tools of Government,* Christopher Hood (first edition 1983, second edition 2007 with Margetts) introduced an analytical approach by developing a theoretical framework to understand what policy instruments – or tools – the state uses in governing. This analytical approach consists of four overarching categories of tools: nodality, authority, treasury and organisation.

In addition to Hood's influential categorisation, there are several later works that summarise and classify various sets of policy instruments. Often, these categorisations are based on the properties of the policy instruments, which can be either hard or soft (measures), push or pull (direction), voluntary or mandatory (force), etc. (Vedung, 2011). Furthermore, policy instruments may range from financial incentives to administrative carrots and sticks, as well as spatial and physical instruments. In the book *Sticks, Carrots and Sermons,* Bemelmans-Videc and Vedung (2011) suggest that sticks correspond to formal regulations, carrots to financial instruments, while sermons correspond to information. In practice, authorities operating at different levels in the public sector have different access to different instruments (Kassim & Le Galès, 2010). For example, the state often uses taxes as a policy instrument, while it is generally cities and municipalities that employ land use as a policy instrument.

The knowledge production about policy instruments can be roughly divided into two camps as well; these camps have their own literatures and they rarely come into dialogue with each other (Howlett, 2005). Firstly, there is the economics literature on policy instruments. This literature is generally concerned with understanding and developing incentive structures and adapting economic or financial policy instruments to either punish or promote certain types of decisions or behaviours within a specified subsector of society or a clearly defined target group. The cost-effectiveness of certain policy instruments has also become a crucial question for economists. Secondly, there is the political science literature on instruments. Instead of studying what policy instruments lead to, given certain assumptions about decisions, behaviours and welfare effects, political scientists are often interested in how policy instruments are used, and what side-effects they cause. Of course, this is an immensely simplified picture of the two camps of knowledge production and their overarching research interests, but it nevertheless shows the basic features of much of the research that has been and still is produced on policy instruments (see discussion in Howlett, 2005).

During the past twenty years, scholars of policy instruments have also turned their attention to policy instrument selection and how policy instruments are becoming mixed in use (Bemelmans-Videc & Vedung, 2011; Bressers & O'Toole, 2005). Many research projects have been inspired by the observation that when multiple instruments are used at the same time, it tends to lead to knock-out effects or encourage decisions or behaviours that move in opposite directions than what was originally intended. As a result, much of the focus in the political science literature has moved on to finding criteria for selecting and designing 'the right' instrument or instrument mixes (see, e.g. Rist, 2011). But the question of the legitimacy of policy instruments has also gained much interest (Wallner, 2008). Research has showed that unless the policy instruments are considered legitimate by the target group or those affected by them, their effect may lead to unintended or even unwanted consequences (Galès, 2010).

While there is critical reflection and critique within these two camps of knowledge production, criticism has also emerged from other disciplines. Political sociologists Lascoumes and Le Galès (2007) have been critical of the assumptions behind much of the political science literature on policy instruments. They summarise the assumptions underlying the research by stating that public policy, in this literature, is primarily conceived as 'pragmatic — that is, as an apolitical and technical approach to solving problems through instruments', which in turn are understood as 'natural', whereby politicians and policymakers understand them as being at their 'disposal', and 'the only questions they raise relate to whether they are the best possible ones for meeting the objectives set' (Lascoumes & Le Galès, 2007, p. 2). Aware of the limitations and unintended consequences of the traditional instruments, scholars have started searching for new instruments. These scholars are generally pragmatic in their approach and they either seek to develop alternative instruments or attempt to design meta-instruments (e.g. characterised by collaboration networks and dialogue), which enable the coordination of the traditional instruments. According to Lascoumes and Le Galès (2007), a focus on different policy networks has emerged, although much of the research on such networks generally converge on a very few instruments that organise and regulate the boundaries and relationships within the networks.

This critique has received much attention and the processual approach to policy instrumentation, promoted by Lascoumes and Le Gales (2007), has been gaining a lot of followers. A processual approach suggests, among other things, that the public authority's choice of instruments and instrument mixes reflects the spirit and values of the times (Stead, 2018). Also, policy instruments are sensitive to trends and political sentiments among the public in that they are influenced both by research results and ideological climate.

The Limits of Policy Instruments

Policy instruments have limits. These limits may be connected to the missing effects of the instrument, or they may be related to the ambiguity of the objective. Either way, the policymakers who designed and selected the policy instrument can rarely control its effects. While the effects may be a result of the intentions behind the policy instrument, they can also have other causes, unknown to the policymakers. One way to think about this gap between intentions and effects is in terms of affordances, a concept that describes how a person, or a group of people, interact with, or respond to, for example, a policy instrument (Faraj & Azad, 2012). While the designer often has certain users and usages in mind, the de facto usage and the user-groups picking up the instrument in question are beyond the designers' control. Understood as the gap between the aims and effects, 'affordances' is a concept that, at least in this context, describes spaces of actions for the target audience.

Thus, policy instruments operate by creating and delimiting action spaces rather than leading to specific, controllable effects. Since those who designed and selected the policy instrument cannot control its effects, the effects should instead be understood as a result of a combination of:

> (i) the properties of the instrument; (ii) the context of the target community to which the instrument is addressed; and (iii) the target community's own propensities to act in particular ways (e.g. given certain desires, beliefs and opportunities that they may have). (Hellström & Jacobs, 2017, p. 605)

Although the context in which the policy instruments are supposed to operate shapes their ability to impact on the target group (Sørensen & Paulsson, 2019, in press), the policy instrument is in itself driving and steering behaviours within the context and may also in fact partly disrupt the structure of the context.

As presented above, the research on policy instruments is rich and the theoretical height impressive. In fact, the vast literature mirrors the many instruments applied by authorities at different levels of governing – be it national, federal, regional or local. One way of conceptualising and also pointing to the potential limits of policy instruments is to discuss multi-level governance and the complexities involved in this. Since authorities at different levels of government are responsible for specific instruments, it might be that different authorities develop and design instruments that shape behavior and investments in opposing directions. For example, a policy instrument that foster transport efficiency might not go hand in hand with an instrument enabling sustainable mobility.

In order to organise and coordinate various policy instruments, collaboration and networks for integrated planning are used as meta-policy instruments, according to Lascoumes and Le Gales (2007). A single instrument cannot change anything on its own, as it were, and that is why a package of instruments is often called for. Since single instruments may only lead to minor changes, the overall effect can become much bigger if the whole policy instrument package is greater than its individual parts (Givoni, 2013). For example, Vieira, Moura, and Viegas (2007) have found that instruments to manage the supply of transport, regulations to influence behavioural patterns and economic instruments to incentivise desired decisions are all instruments that support each other and where synergies have been identified.

As mentioned earlier, policy instruments may come in conflict with each other and thereby limit their potential for governing. For example, the effects caused by one instrument may be reversed by another, making the policy package a zero-sum game. As also mentioned before, different instruments are not only good for different things, they also differ when it comes to cost-effectiveness. Congestion charges require costly asset-specific infrastructures and data management capabilities, while reduced parking space is a relatively inexpensive measure. Sometimes certain regulations are very costly for the target group, although the effects may be marginal. According to this way of reasoning, a policy instrument, whose effect is limited in relation to its cost, should probably not be implemented (cf. Frey, 2003). While cost-effective instruments with great intended effects are often the official ambition of policymakers, it is at the same time difficult to know in advance what the effects will be. Affordances conceptualise this ambiguity of the action space elicited by a policy instrument. Pilots, test beds, policy labs or other forms of small-scale trials are often implemented in order for researchers and governments to evaluate the effect, or rather the affordances, of the policy instrument before it is introduced on a large scale.

Outline of the Book

Apart from this introductory chapter, the book is organised into three parts, each including a number of chapters, as well as a final chapter with concluding remarks.

The following part includes chapters explaining *why* there is a need for policy instruments in relation to smart mobility. This part begins with Chapter 2 by

Moscholidou, which explores how officials from English transport authorities see state intervention evolve in the future, and what accountability arrangements are necessary to achieve the level of steering they envisage. The chapter shows that the officials all expect new services to improve the local transport provision. The chapter closes with a call to rebalance the narrative around smart mobility as well as focussed action. Chapter 3 in this part is authored by Pernestål, Engholm, Kristoffersson, and Hammes and relates to how CLDs (System Dynamics and Causal Loop Diagrams) can be used to describe the effects of automation of the transport system, and relationships between different effects as well as the potential impact of policies. One insight is that policies are needed to balance the decreased marginal cost of road vehicles to avoid greater energy consumption and increased emissions.

In the second part, the authors attempt to discuss the processes of *how* policy instruments may be chosen and developed. In these chapters, the focus is on how expert knowledge, multi-stakeholder contexts and democratic processes potentially contribute to shaping policy instruments. Docherty opens this section in Chapter 4 by asking whether it is easier within multi-level governance structures to implement new policy instruments or vice versa. In the chapter, he explores the key challenges that some potential policy instruments such as new pricing models and road space management will face in being delivered in multi-level governance systems. In Chapter 5, McLeod, Curtis and Stone examine the extent to which studies of transport modelling contribute to 'smart' mobility knowledge. They suggest that caution is needed in planning for potential futures based on the limited existing knowledge base and assert a need for planning support tools that can offer more pluralistic, discursive and transparent methods. In Chapter 6, Stone, Ashmore, Legacy and Curtis take their starting point in Australia and elicit the opinions of key players in the public and private sectors on the scope of public regulation required to manage technology transitions in urban transport. The ambition of the authors is to help focus attention on which forms of regulation might be supported by industry and government, and whether these would be enough to manage complex transitions. Kronsell and Mukhtar-Landgren in Chapter 7 of this part focus on local experiments including public and private actors as a policy instrument to promote smart mobility solutions. They offer a conceptual analysis of how the diverse roles that municipalities take on in experimental governance relate to democratic concerns for inclusion, legitimacy, power and transparency.

The chapters in the third part discuss *what* policy instruments are doing and what smart mobility is doing to them. In Chapter 8, Reardon discusses the extent to which smart mobility can help create policy change towards the goal of low carbon mobility. She suggests that smart mobility innovations may make it possible to envisage smart mobility by incrementally changing policy and policy instruments towards low carbon mobility. Looking into the large amount of policy instruments available for the state, Wallsten, Sørensen, Paulsson and Hultén in Chapter 9 analyse the governing capacity of national policy instruments and how they might be affected in two different scenarios of smart mobility. They conclude that many policy instruments still possess governing capacity, for example, instrument-related authority and treasure, while governing capacity might be reduced for instruments which are based on the state's nodality or organisation. Following this, in Chapter 10, Fearnley focusses on micromobility, namely e-scooters. He describes the developments and regulatory dilemmas, and discusses new and innovative – and potentially efficient – methods of regulation based on international experiences. Taking their point of departure in a Swedish context, Pettersson-Löfstedt and Khan in Chapter 11 focus on public transport in rural areas and explore the conditions and challenges connected to ICT and autonomous vehicles. They discuss how governance needs to change to increase attention to rural areas and provide suggestions of policy instruments that can be applied.

In the final and concluding chapter, Sørensen and Paulsson provide overall conclusions and look ahead. They do so by answering the why, how and what of policy instruments for shaping smart mobility futures. They furthermore discuss a transition to a sustainable society reaching climate reduction targets and suggest that instruments establishing a much-needed dialogue with and involvement of citizens in sustainability transitions need to be studied and developed.

Conclusions

The overarching ambition of this volume is to explore which policy instruments that are used to govern smart mobility futures towards sustainability transitions. Policy instruments are defined in this book broadly as mechanisms of governance that, in one way or another, involve the use of public authority.

We started this introductory chapter by briefly discussing the intricate relationships between societal objectives and policy instruments before moving on to discuss the literature on policy instruments. There is extensive literature on policy instruments and a plethora of categorisations and perspectives, ranging from rationalist, instrumentalist to interpretative and processual. In the transport sector alone, the literature on policy instruments is massive. There is also an emerging body of literature on the governance, planning and policies of smart mobility. Much of this research lands in the conclusion that political involvement is desirable to some extent, not least if societal objectives are going to be achieved as part of this ongoing socio-technological transition, but relatively limited attention has so far been devoted to how current policy instruments are impacted or how new instruments are to be designed to realise societal objectives in an era of smart mobility.

The chapters in this book provide an insight into how this can happen. By going beyond the statement that there is a need of governance, the chapters point to the instruments used today and how they might have to be calibrated in futures of smart mobility.

Acknowledgements

We owe all the chapter authors in this book a big thank you. Without their constructive comments at a workshop in Lund on 25–26 September 2019, this book would not have taken the shape it did. Louise Reardon and Iain Docherty have provided valuable input on how to sharpen the structure and content of the chapter. We are also grateful for research funding from three organisations: the Swedish Energy Agency, the Swedish Innovation Agency (Vinnova), and the Swedish Knowledge Centre for Public Transport Research (K2).

References

Ali, S., Hussain, T., Zhang, G., Nurunnabi, M., & Li, B. (2018). The implementation of sustainable development goals in "BRICS" countries. *Sustainability*, *10*, 2513–2527. doi:10.3390/su10072513

Bemelmans-Videc, B.-M., & Vedung, E. (2011). Policy instruments types, packages, choices, and evaluation. In M. L. Bemelmans-Videc, R. C. Rist, & E. O. Vedung (Eds.), *Carrots, sticks, and sermons: Policy instruments and their evaluation*. (pp. 249–274) London: Transaction Publishers.

Bjelfvenstam, J. (2018). Vägen till självkörande fordon [The Road To Self-Driving Vehicles] (SOU 2018:16). Retrieved from https://www.sou.gov.se/?s=2018%3A16&limiter=allt. Accessed on December 2, 2019.

Bressers, H. Th. A., & O'Toole L. J. Jr. (2005). Instrument selection and implementation in a networked context. In P. Eliades, M. M. Hill, & M. Howlett (Eds.), *Designing government: From instruments to governance* (pp 132–153) . London: McGill-Quenn's University Press.

C40. (2019). About C40. Retrieved from https://www.c40.org/about. Accessed on September 16, 2019.

COWI & PTV. (2019). *The Oslo study – How autonomous cars may change transport in cities*. Oslo: Ruter. Retrieved from http://www.mynewsdesk.com/no/ruter/documents/the-oslo-study-how-autonomous-cars-may-change-transport-in-cities-87313. Accessed on September 16, 2019.

Dickinson, J., & Wretstrand, A. (2015). Att styra mot ökad kollektivtrafikandel. [Governing towards an increasing share of public transportation] K2 Research 2015:2, K2, Lund. Retrieved from http://www.k2centrum.se/sites/default/files/fields/field_uppladdad_rapport/styrmedel_k2research.pdf. Accessed on November 27, 2019.

Docherty, I., Marsden, G., & Anable, J. (2018). The governance of smart mobility. *Transportation Research Part A*, *115*, 114–125.

Dowling, R. (2018), Smart mobility: Disrupting transport governance?. In G. Marsden & L. Reardon (Eds.), *Governance of the smart mobility transition* (pp. 51–64). Bingley: Emerald Publishing Limited. https://doi.org/10.1108/978-1-78754-317-120181004

Edmondson, D. L., Kern, F., & Rogge, K. S. (2018). The co-evolution of policy mixes and socio-technical systems: Towards a conceptual framework of policy mix feedback in sustainability transitions. *Research Policy*, *48*(10), 103555. http://doi.org/https://doi.org/10.1016/j.respol.2018.03.010

Fagnant, D. J., & Kockelman, K. M. (2015). Preparing a nation for autonomous vehicles: Opportunities, barriers and policy recommendations. *Transportation Research Part A: Policy and Practice*, *77*, 167–181.

Faraj, S., & Azad, B. (2012). The materiality of technology: An affordance perspective. In P. M. Leonardi, B. A. Nardi, & J. Kallinikos (Eds.), *Materiality and organizing: Social interaction in a technological world*.

Finger, M., & Audouin, M. (Eds.). (2018). *The governance of smart transportation systems*. Berlin: Springer.

Frey, B. (2003). Why are efficient transport policy instruments so seldom used?. In J. Schade & B. Schlag (Eds.), *Acceptability of transport pricing strategies* (pp. 63–75). Bingley: Emerald Group Publishing Limited. https://doi.org/10.1108/9781786359506-004

Galès, P. (2010). Policy instruments and governance. In M. Bevir (Ed.), *Handbook of governance* (pp. 142–159). Thousand Oaks, CA: Sage. http://doi.org/10.4135/9781446200964.n10

Givoni, M. (2014). Addressing transport policy challenges through policy-packaging. *Transportation Research Part A*, *60*, 1–8.

Gössling, S., & Cohen, S. (2014). Why sustainable transport policies will fail: EU climate policy in the light of transport taboos. *Journal of Transport Geography*, *39*, 197–207. https://doi.org/10.1016/j.jtrangeo.2014.07.010

Grazi, F., & van den Bergh, J, C. J. M. (2008). Spatial organization, transport, and climate change: Comparing instruments of spatial planning and policy. *Ecological Economics*, *67*(4), 630–639. http://doi.org/https://doi.org/10.1016/j.ecolecon.2008.01.014

Hellström, T., & Jacob, M. (2017). Policy instrument affordances: A framework for analysis, *Policy Studies*, *38*(6), 604–621. doi:10.1080/01442872.2017.1386442

Herrmann, A., Brenner, W. M., & Stadler, R. (2018). *Autonomous driving. How the driverless revolution will change the world.* Bingley: Emerald Publishing Limited.

Hildebrandt, S. (Ed.). (2016). *Bæredygtig global udvikling. FN's 17 verdensmål i et dansk perspektiv* [Sustainable Global Development: UN's 17 SDGs in a Danish Perspective]. København: Jurist- og Økonomforbundets Forlag.

Hood, C. C., & Margetts, H. Z. (2007). *The tools of government in the digital age.* London: Palgrave Macmillan.

Hopkins, D., & Schwanen, T. (2018). Governing the race to automation, In G. Marsden & L. Reardon (Ed.), *Governance of the smart mobility transition* (pp. 65–84). Bingley: Emerald Publishing Limited. https://doi.org/10.1108/978-1-78754-317-120181005

Howlett, M. (2005). What is a policy tool? Policy tools, policy mixes and policy implementation styles. In P. Eliadis, M. Hill, & M. Howlett (Eds.), *Designing government: From instruments to governance* (pp. 31–50). London: McGill-Queen's University Press.

Howlett, M. (2009). Governance modes, policy regimes and operational plans: A multi-level nested model of policy instrument choice and policy design. *Policy Sciences*, *42*, 73–89. https://doi.org/10.1007/s11077-009-9079-1

Howlett, M., & Ramesh, M. (1993). Patterns of policy instrument choice: Policy styles, policy learning and the privatization experience. *Review of Policy Research*, *12*(1–2), 3–24. http://doi.org/10.1111/j.1541-1338.1993.tb00505.x

Hrelja, R., Isaksson, K., & Richardson, T. (2013). Choosing conflict on the road to sustainable mobility: A risky strategy for breaking path dependency in urban policy making. *Transportation Research Part A: Policy and Practice*, *49*, 195–205. https://doi.org/10.1016/j.tra.2013.01.029

ICLEI. (2019). About us. Retrieved from https://www.iclei.org/en/About_ICLEI_2.html. Accessed on September 16, 2019.

Institute for Transport Studies. (2009). Policy instruments: A policy guidebook. Retrieved from http://www.its.leeds.ac.uk/projects/konsult/public/level1/sec09/index.htm. Accessed on December 2, 2019.

International Transport Forum. (2017a). *Transition to shared mobility. How large cities can deliver inclusive transport services.* Paris: OECD. Retrieved from https://www.itf-oecd.org/transition-shared-mobility. Accessed on September 16, 2019.

International Transport Forum. (2017b). *Shared mobility simulations for Auckland.* Paris: OECD. Retrieved from https://www.itf-oecd.org/shared-mobility-simulations-auckland. Accessed on September 16, 2019.

International Transport Forum. (2017C). *Shared mobility simulations for Helsinki.* Paris: OECD. Retrieved from https://www.itf-oecd.org/shared-mobility-simulations-helsinki. Accessed on September 16, 2019.

International Transport Forum. (2018). *Shared mobility simulations for Dublin.* Paris: OECD. Retrieved from https://www.itf-oecd.org/shared-mobility-dublin. Accessed on September 16, 2019.

Janecki, R. (2011). Intelligent transportation systems in transportation policy of the cities. In J. Mikulski (Eds.), *Modern transport telematics* (pp. 265–276). TST 2011. Communications in Computer and Information Science 239. Heidelberg: Springer.

Kassim, H., & Le Galès, P. (2010). Exploring governance in a multi-level polity: A policy instruments approach. *West European Politics*, *33*(1), 1–21. http://doi.org/10.1080/01402380903354031

Lascoumes, P., & Le Gales, P. (2007). Understanding public policy through its instruments — From the nature of instruments to the sociology of public policy instrumentation. *Governance*, *20*(1), 1–21. http://doi.org/10.1111/j.1468-0491.2007.00342.x

Lenton, T. M., Rockström, J., Gaffney, O., Rahmstorf, S., Richardson, K., Steffen, W., & Schellnhuber, H. J. (2019). Climate tipping points — Too risky to bet against. *Nature*, *575*, 592–595.

Lyons, G. (2018). Getting smart about urban mobility – Aligning the paradigms of smart and sustainable. *Transportation Research Part A*, *115*, 4–14.

Marsden, G., & Reardon, L. (Eds.). (2018). *Governance of the smart mobility transition*. Bingley: Emerald Publishing Limited.

Nash, C., & Matthews, B. (2013). Transport pricing and subsidy. In J. Rodrigue, T. Notteboom, & J. Shaw (Eds.), *The SAGE handbook of transport studies* (pp. 293–310). London: SAGE Publications, Ltd. doi:10.4135/9781446247655.n17

Notteboom, T. (2013). Transport policy instruments. In J. Rodrigue, T. Notteboom, & J. Shaw (Eds.), *The SAGE handbook of transport studies* (pp. 281–292). London: SAGE Publications, Ltd. doi:10.4135/9781446247655.n16

Papa, E., & Lauwers, D. (2015). Smart mobility: Opportunity or threat to innovate places and cities. In *Proceedings of 20th international conference on urban planning and regional development in the information society* (pp. 543–550), University of Westminster, Ghent, Belgium.

Petterson-Löfstedt, F., & Sørensen, C. H. (2019, in press). Why do cities invest in bus priority measures? Policy, polity, and politics in Stockholm and Copenhagen, *Transport Policy*, in press. https://doi.org/10.1016/j.tranpol.2019.10.013

Pietzsch, M. (2018). Demand ridesharing as a part of local public transport. The way to a car-free city. Presentation at Kollekivforum, Oslo, February 5, 2018. Retrieved from http://kollektivforum.no/getfile.php/1347042/Kollektivforum/Michael%20Pietzsch.pdf. Accessed on September 16, 2019.

Reardon, L., & Marsden, G. (2018). Conclusions: A window of opportunity. In G. Marsden & L. Reardon (Eds.), *Governance of the smart mobility transition* (pp. 155–165). Bingley: Emerald Publishing Limited.

Rist, R. C. (2011). Choosing the right policy instrument at the right time: The contextual challenges of selection and implementation. In M. L. Bemelmans-Videc, R. C. Rist, & E. O. Vedung (Eds.), *Carrots, sticks, and sermons: Policy instruments and their evaluation* (pp. 149–164). London: Transaction Publishers.

Rodrigue, J. (2013). Urban transportation and land use. In J. Rodrigue, T. Notteboom, & J. Shaw (Eds.), *The SAGE handbook of transport studies* (pp. 105–118). London: SAGE Publications, Ltd. doi:10.4135/9781446247655.n7

Rogge, K. S., & Reichardt, K. (2016). Policy mixes for sustainability transitions: An extended concept and framework for analysis. *Research Policy*, *45*(8), 1620–1635. http://doi.org/https://doi.org/10.1016/j.respol.2016.04.004

Santos, G., Behrendt, H., & Teytelboym, A. (2010). Part II: Policy instruments for sustainable road transport. *Research in Transportation Economics*, *28*(1), 46–91. http://doi.org/https://doi.org/10.1016/j.retrec.2010.03.002

Schiller, P. L. (2016). Automated and connected vehicles: High tech hope or hype. *World Transport Policy and Practice*, *22*(3), 28–44.

Seba, T. (2014). *Clean disruption of energy and transportation. How Silicon Valley will make oil, nuclear, natural gas, coal, electric utilities and conventional gas obsolete by 2030*. Silicon Valley: Clean Planet Ventures.

Sørensen, C. H., & Gudmundsson, H. (2010). Målstyret transportpolitik – Hvad kan Danmark lære af Norge og Sverige? [Management by objectives in transport policy - What can Denmark Learn from Norway and Sweden?] *Økonomi & Politik*, *83*(2), 3–19.

Sørensen, C. H., & Paulsson, A. (2019, in press). Contextualizing policy: Understanding implementation under socio-technical transitions. *International Journal of Public Administration*, in press. https://doi.org/10.1080/01900692.2019.1665067

Stead, D. (2018) Policy preferences and the diversity of instrument choice for mitigating climate change impacts in the transport sector. *Journal of Environmental Planning and Management*, *61*(14), 2445–2467. doi:10.1080/09640568.2017.1397505

Stead, D., & Vaddadi, B. (2019). Automated vehicles and how they may affect urban form: A review of recent scenario studies. *Cities*, *92*, 125–133. http://doi.org/https://doi.org/10.1016/j.cities.2019.03.020

United Nations (2019). About the sustainable development goals. Retrieved from https://www.un.org/sustainabledevelopment/sustainable-development-goals/. Accessed on September 16, 2019.

Vedung, E. (2011). Policy instruments: Typologies and theories. In M. L. Bemelmans-Videc, R. C. Rist, & E. O. Vedung (Eds.), *Carrots, sticks, and sermons: Policy instruments and their evaluation* (pp. 21–58). London: Transaction Publishers.

Vieira, J., Moura, F., & M. Viegas, J. (2007). Transport policy and environmental impacts: The importance of multi-instrumentality in policy integration. *Transport Policy*, *14*(5), 421–432. https://doi.org/10.1016/j.tranpol.2007.04.007

Wallner, J. (2008). Legitimacy and public policy: Seeing beyond effectiveness, efficiency, and performance. *Policy Studies Journal*, *36*(3), 421–443. http://doi.org/10.1111/j.1541-0072.2008.00275.x

Part I

Chapter 2

Steering Smart Mobility Services: Governance and Accountability Challenges for English Local Authorities

Ioanna Moscholidou

ABSTRACT

There are different narratives surrounding smart mobility, which can sometimes even appear as opposing (Lyons, 2018). Its fiercest proponents are promising versions of a revolutionised future, where users have on-demand access to multiple mobility options and are freed from car ownership, while transport systems become carbon neutral and congestion is a problem of a bygone age (Sherman, 2019). At the same time, the plausibility of such visions of the future has been questioned, with critics warning against the potentially negative impacts of the widespread adoption of privately provided services and stressing the need for state intervention to avoid exacerbating 'classic' transport issues such as congestion and unequal access to services, as well as creating new challenges such as uncontrolled market monopolies (Docherty, Marsden, & Anable, 2018). Drawing from these narratives, this chapter explores how officials from English transport authorities see state intervention evolve in the future, and what accountability arrangements are necessary to achieve the level of steering they envisage. Based on interviews with local authority officials, this chapter shows that the officials' views generally align more closely with the narrative of providers than with that of critics. Although different local authorities envisage varying levels of control and steering of smart mobility, they all expect new services to improve the local transport provision. This chapter also discusses the barriers local authorities face in shaping local accountability arrangements.

Keywords: Smart mobility; governance; accountability regimes; local transport policy; English transport authorities; sustainable mobility

Shaping Smart Mobility Futures:
Governance and Policy Instruments in times of Sustainability Transitions, 19–35

doi:10.1108/978-1-83982-650-420201002

Potential Benefits and Risks of Smart Mobility

Smart mobility service providers often promise a revolutionised future, where people perform mobility in a seamless, on-demand and sustainable way. For example, much of the discussion around dockless bikes and scooters highlights their potential to cover the first and last mile of multi-modal trips. Similarly, ridehailing platforms emphasise their potential to connect people to public transport nodes and scale up sharing to the extent that people no longer need a car; while Mobility-as-a-Service (MaaS) applications promise seamless and integrated access to a wide range of daily activities through on-demand mobility. This rhetoric is certainly catching the attention of cities, and there are already multiple examples of trials, policies and plans for smart mobility futures. However, critics warn about the complex issues that may arise from the unexamined adoption of new services and highlight the need for timely state intervention to steer smart mobility towards a direction that delivers public value (among others see Docherty, Marsden, & Anable, 2018; Lyons, 2018; Marsden & Reardon, 2018).

Acknowledging the complex balance, the state needs to strike between delivering public value and not stifling innovation, Docherty et al. (2018) outline a range of challenging issues that may arise in a smart mobility transition and require the intervention of the state. These include, but are not limited to, the changing set of actors in the mobility arena who challenge traditional business models and the way the state interacts with the private sector; the need to change taxation structures to reflect the changing car ownership models and the decreasing reliance on fossil fuels; as well as issues of data sharing, equity and inclusion and balancing profitable operation of services with social obligation (Docherty et al., 2018). Echoing Docherty et al. (2018), Lyons (2018) stresses the importance of aligning the paradigms of smart and sustainable and highlights existing dichotomous discourses on either side. He explains that interpretations of smart and the goals pursued through smart mobility do not necessarily align when used by technologists and urban planners. Large corporations are exerting significant influence in the 'era of smart' based on principles that are conflicting with traditional concerns of the public sector such as social and environmental sustainability as well as economic prosperity. In order to address the conflict in the different interpretations, Lyons (2018) comes up with a new definition for smart urban mobility that brings forward public policy principles: *connectivity in towns and cities that is affordable, effective, attractive and sustainable*.

Further research highlights the risks associated with specific types of smart mobility services. For example, Pangbourne, Mladenović, Stead, and Milakis (2019) explore whether promises for seamless and sustainable mobility made by MaaS providers have unintended consequences, and they identify the governance challenges associated with these. They find that MaaS has considerable potential for deception. As profitability for private businesses focusses on growth in the use of their services, there is potential for increased mobility among those who can pay for it, resulting in the exacerbation of issues such as poor air quality and congestion, while still not helping those who experience transport poverty. They highlight that if platforms are to influence behaviours to be more sustainable, this must be designed in from the start, particularly as the bundled mobility pricing may obscure the true costs of individual journeys (Pangbourne et al., 2019).

The literature also highlights the potential benefits of smart mobility. Although smart mobility services will not necessarily help deliver public policy objectives if not steered, they also present transport authorities with an unprecedented opportunity to address long-standing problems through rapidly advancing technological innovation. These problems are not limited to the type and quality of mobility that is currently available, but also to how transport is governed. As Marsden and Docherty (2019) point out:

> the smart mobility transition is an opportunity to remove or improve some of the challenges that the current structures create and to achieve goals which have proven to be difficult with the existing technologies, structures and incentives. (Marsden & Docherty, 2019, p. 4)

However, they also stress that if smart mobility is to be approached as a means to address existing problems as well as a way to overcome local challenges, it is necessary to understand how this can be achieved. In this book, Chapter 8 by Reardon (2020) highlights that when smart mobility innovations are approached as policy instruments, it is possible to envisage smart mobility incrementally changing policy towards low carbon mobility under certain conditions.

Steering Smart Mobility and Local Accountability Arrangements

The literature is clear about the potential risks of smart mobility and the need for proactive state intervention. But how do local and national governments approach these risks and what do they do to steer smart mobility? Legacy, Ashmore, Scheurer, Stone, and Curtis (2019) explore these questions in the Australian context and investigate the role that urban planning can play in the rollout of Autonomous Vehicles (AVs) in order to anticipate and mediate unwanted impacts, such as exacerbating unequal access to transport, further contributing to the climate emergency and increasing the fragmentation of physical landscapes. Through interviews with planning and regulatory agencies across the Australian public sector, they find that although planning officials were aware of the need to shape the AV future in ways that meet public policy goals, there was little understanding of how this could be achieved (Legacy et al., 2019).

This chapter aims to build on this body of literature and explores how local authorities in England currently steer smart mobility services and how they expect their position to evolve in the future. For the purposes of this research, steering is narrowed down to what the state does to hold smart mobility providers accountable for their actions. This framing is particularly useful for exploring arrangements at the local level as steering can also involve, for example, state-led research and development and the development of a broader innovation agenda at the national level. However, this chapter does not take into account the perspective and views of the private sector, including smart mobility providers themselves. This is a crucial part of the debate that should not be underestimated. In this book's Chapter 6, Stone, Ashmore, Legacy, and Curtis (2020) explore the views of both public and private sector actors in the Australian context, drawing important conclusions on the types of regulations that may be appropriate for AVs.

To explore accountability arrangements, this chapter uses Mashaw's (2006) accountability regimes as a framework. Mashaw (2006) suggests that any talk about accountability is at its base talk about six questions, which form the building blocks of what he calls 'accountability regimes'. Accountability is 'unpacked' through six questions: who is liable or accountable and to whom; what are they liable to be called to account for; through what processes accountability is to be assured; by what standards the putatively accountable behaviour is to be judged; and what the potential effects are of finding that those standards have been breached (Mashaw, 2006, p. 118). This framework is used to develop a general picture of the steering approach that local authorities take towards smart mobility through exploring a broad range of accountability aspects such as the relationship between the state and smart mobility providers (who is accountable and to whom), as well as what purposes smart mobility is expected to serve (for what) and the policy tools through which steering is achieved (processes, standards and effects). As such, this chapter does not provide an in-depth exploration of each local context.

Using this framing, this chapter specifically asks the following questions:

- What are the current accountability arrangements in the local authorities studied?
- What role do local authorities see themselves playing in steering smart mobility in the future and what are the accountability arrangements needed to achieve this?
- What are the barriers that local authorities are facing in creating the accountability arrangements that support their role in the future?

Local Transport Governance in England

This section provides a brief overview of local transport governance arrangements in the United Kingdom and England in order to set into context the analysis in the following sections. The governance of the transport system in the United Kingdom is complex and uneven. Government intervention in transport has varied across time, mode and region and has historically played a role in creating this complexity. Since its inauguration in 1919, the Ministry of Transport tended to be reactive rather than agenda-setting, and followed a clear predict-and-provide approach in transport planning at least until the 1990s. Central government tended to treat modes separately, effectively preventing the development of an integrated transport policy for the United Kingdom as a whole (Gunn, 2018). A key central government intervention that still affects local public transport was the deregulation of the bus system through the Transport Act 1985, which largely abolished any regulation related to bus provision (routes and frequency) and price outside London and Northern Ireland (Marsden & Docherty, 2019).

Over the recent decades, centralised control of transport has been gradually and asymmetrically devolved. Following respective devolution acts in 1998, Scotland, Wales and Northern Ireland all possess executive and legislative powers over the strategic planning of transport, including roads. Further devolution within England saw the creation of the Greater London Authority and the Mayor of London, and later Combined Authorities and Metro Mayors outside London. Devolved powers vary across the country depending on the agreements Combined Authorities have individually reached with central government. Metro Mayors in parts of England and the Mayor of London have only executive powers. Combined Authorities and the London Assembly[1] can scrutinise executive decisions but not legislate them in the manner of the Scottish Parliament, National Assembly for Wales and the Northern Ireland Assembly (Torrance, 2019).

This chapter focusses on English regional and local transport authorities outside London, which was deliberately left out as its governance arrangements and institutional powers over planning, delivering and operating the transport system are very different to the rest of the country. Local authorities outside London have fewer powers and deal with a significantly less sustainable transport mix, which creates a relatively even basis for comparing their thinking about future mobility in this chapter.

Outside London, authorities whose remit includes transport can exist at up to six spatial tiers (national, pan-regional, regional, sub-regional, local and district). Over the years, there has been a trend towards layering of institutions that has also added layers to decision-making processes. This sometimes results in misaligned and overlapping boundaries between authorities but has also brought transport closer to other policy areas, such as employment and skills (Marsden & Docherty, 2019). Transport functions vary across different institutional structures. At the local level, large town and city councils have a transport function that allows them to manage local roads and traffic, as well as develop long-term local transport plans and determine local policies, such as parking. Internal structures vary across different authorities but, in general, transport departments within the councils of bigger cities tend to be more independent (as opposed to merged with other departments, such as planning) and have more specialised posts. Specialised posts may focus on transport innovation, air quality management, walking and cycling, etc.

At the regional level, Combined Authorities are set up across two or more neighbouring councils aiming to coordinate responsibilities and powers over certain services, including aspects of transport, housing and social care. Some Combined Authorities have executive transport bodies, which are often the evolution of pre-existing regional transport authorities that were rebranded and were given

[1]The Greater London Authority comprises the London Assembly and the Mayor of London.

more powers after the creation of the Combined Authorities. Such executive bodies are Transport for West Midlands (TfWM) and Transport for Greater Manchester (TfGM), which are responsible for developing the integrated transport strategy and policies for the West Midlands Combined Authority and the Greater Manchester Combined Authority, respectively. Other Combined Authorities, such as the West Yorkshire Combined Authority (WYCA), maintain their transport function in house. Combined Authorities have more extensive powers than the local councils that comprise them, including some franchising powers over local public transport, and therefore their internal structures generally include more specialist posts.

Participant Selection and Interviews

The transport authorities approached for this research already had some interaction with smart mobility providers or were already supportive of local trials. Engagement between local transport authorities and smart mobility providers is not consistently carried out through a dedicated department or officials within each local authority. When there is an innovation team or post within the transport authority, it is more likely that engagement will be coordinated by or targeted towards them but the extensively disruptive character of smart mobility means that multiple teams within local transport authorities may interact with providers. Therefore, the participants contacted were senior transport innovation or strategy officials who were more likely to have a broad understanding of how each transport authority deals with smart mobility providers. All invitees had direct involvement in shaping their organisation's response to smart mobility services through the development of trials, strategies and policies. Invitations to participate were sent out to officials from seven local authorities, which were deliberately chosen to be diverse in their remit and smart mobility initiatives. Five of the seven officials agreed to participate, representing three combined authorities (Greater Manchester, West Midlands and West Yorkshire) and two city councils (Leicester and Milton Keynes). Although this sample is not considered representative of all perspectives across the country, it is considered sufficient to develop an early insight into what role local authorities see themselves playing in steering smart mobility and their views on present and future accountability arrangements.

Five semi-structured interviews were carried out, each lasting an hour. In order to frame the discussion, at the beginning of each interview participants were provided with three 'accountability scenarios'. Each scenario represented a city's[2] approach towards steering smart mobility and was linked to three different descriptions of accountability arrangements for smart mobility providers. The scenarios were not aimed to be fully representative of an authority's approach at a moment in time, but a description of its general position. The scenarios ranged

[2]The term city was used for convenience, but each discussion was focussed on each transport authority's actual remit (i.e. regions when it came to TfWM, WYCA, and TfGM).

from laissez-faire with minimal state intervention (City A), to moderate state intervention (City B), to comprehensive intervention (City C). In line with the framing presented in Section 2, for each scenario the accountability arrangements were described using Mashaw's (2006) six-question framework of accountability regimes, as summarised in Table 1. At the start of their interviews, all participants were asked which scenario better describes their organisation's approach presently, where they aim to be in the future, and what are the barriers in getting there. The use of this simple framework introduced into the discussion a range of future challenges for an open-ended timeline and helped participants conceptualise key aspects of accountability.

Smart Mobility Developments in England

Considering that the smart mobility landscape is changing rapidly across the world, elements of this chapter are expected to be out of date at the time of publication. Therefore, this section provides a snapshot of the developments that had already taken place at the time when the interviews were carried out (Autumn, 2018) and at the time of writing (a year later), to help the readers place the discussion in context. Overall, although many services and trials were already in place, no specific bold moves in terms of steering smart mobility can be identified at the national or local level.

At the national level, the government had launched the 'Future of Mobility Grand Challenge' as a key element of the country's Industrial Strategy, aiming to help the United Kingdom become a world leader in shaping the future of mobility. At the time of writing, as part of the challenge the Department for Transport (DfT) had launched a call for evidence about the issues surrounding the future of mobility; had published the 'Future of Mobility: Urban Strategy', which outlines the key principles for steering smart mobility in the future based on the evidence reviewed; had provided funding for Future Mobility Zones,[3] where trials and demonstrations can take place; and had announced a regulatory review focussed on smart mobility. At the time the interviews were carried out only some of the funding for trials had been committed to TfWM and the call for evidence was still open. As such, the discussion in this chapter refers to the period before the DfT had published a smart mobility strategy.

At the local level, some relatively high-profile events related to smart mobility had already taken place. The conflict between Uber and Transport for London (TfL) was well underway with Uber operating with a temporary license as TfL had refused its long-term renewal (see Dudley, Banister, & Schwanen, 2017; Topham, 2018). Also, some dockless bike operators had already launched and started scaling down their operations in a number of cities across the country (Kollewe & McIntyre, 2019). Focussing on the local authorities interviewed, TfWM was already carrying out some trials on AVs, while Birmingham (the biggest city in the West Midlands)

[3]£90m to be allocated to up to four local/regional authorities for Future Mobility Zones.

Table 1. Future Accountability Scenarios Used in the Interviews.

	City A	City B	City C
Description	The city is open to smart mobility services and regulatory arrangements are adjusted so that it becomes easier for new technologies to thrive. There are no specific regulations that apply to smart mobility providers, and new services can compete freely with existing public transport provision. The market is self-regulated through competition but the city still covers 'market failures'	The city is open to smart mobility services but may regulate or restrict them, if necessary, on a case by case basis. Most services need to comply with minimum standards of safety, data sharing, and customer service. On occasion, the city may develop pilot projects or partner with smart mobility providers. The market is self-regulated through competition but the city still covers 'market failures'	The city expects smart mobility services to be integrated with 'traditional' public transport services. Services are procured by the local authority through new collaborative arrangements or permit systems that allow greater flexibility than traditional procurement routes. Smart mobility services fill specific gaps in provision
To whom are smart mobility providers accountable?	Smart mobility providers may be accountable to national, regional or local authorities for complying with baseline standards, such as safety. Services or providers are also held accountable by their users for the level of service they offer	Smart mobility providers may be accountable to national, regional or local authorities for complying with various standards, such as safety and customer service standards, but rules may overlap or not align across different governance levels. Services or providers are also accountable by their users for the level of service they offer	Smart mobility providers are responsible to national, regional and local authorities for standards that are set as a result of coordinated action across different governance levels. Services or providers are also accountable by their users for the level of service they offer

What are they accountable for?	Services are generally accountable only for meeting basic safety and operational standards and baseline regulations, such as data protection	Providers may be accountable for complying with voluntary agreements or meeting the terms and conditions of pilots. It is expected that providers maintain a collaborative relationship with local authorities	Providers are accountable for meeting the terms of their procurement or permits, as well as demonstrating that their operations are contributing towards achieving the city's long-term goals
What are the processes through which providers are held accountable?	Providers are held accountable through national, regional, or local legal and regulatory processes that usually do not apply only to smart mobility services	Providers are held accountable through some processes that are specific to smart mobility services, such as data sharing requirements. Rules may be developed in consultation with providers can be enforced by the local transport authority. Specific collaborative efforts come with their own rules and processes that are set and upheld by all partners involved	Providers are held accountable through the terms and conditions of the contracts as well as overarching rules that create a level playing field for different types of smart mobility services. The development of policies and regulations is led by each responsible authority, and there are clear routes for providers to contest standards or provide their feedback in the process
What are the standards that need to be met?	Standards mostly relate to safety, taxes and customer rights	Standards may include safety, customer services standards and data sharing but many of them are set on a voluntary basis	There are clear standards related to operations and customer service, as well as overarching standards related to the kind of trips that are replaced and targets related to modal shift from the private car to shared smart mobility services
What are the consequences if standards are not met?	Legal and financial implications and limited as the market is considered self-regulated	There are some legal and financial implications, which are often contested by the providers. As many rules are voluntary to comply with, there are few implications when there are not adhered to	There are clear legal and financial implications, as well as loss of contracts and right to develop further partnerships with the city if providers do not comply with regulations

had become the second city where MaaS Global launched its 'whim' application as a MaaS solution. In Milton Keynes, multiple trials and demonstrations were underway, including a docked shared bike system, autonomous pods, autonomous cars and autonomous food delivery robots. In Leeds and Leicester, the local authorities were in discussions with dockless bike providers about their launch, although some of these providers had already left the United Kingdom at the time of the interviews. Finally, Mobike had launched and withdrawn its dockless bikes from Manchester, while TfGM was involved in a number of European and national projects focussing of different aspects of MaaS and AVs.

Findings and Discussion

Current Accountability Regimes and Uncertainty

When discussing the current accountability regime, participants' views were broadly aligned, which is not unexpected as regulations and powers over transport are almost identical across the authorities interviewed. All participants identified their current position as between City A and City B, stating that they have some ability to control and steer smart mobility but most of its aspects are controlled by the market. The interviews revealed that present accountability arrangements are partly defined by policy tools and regulations that predate smart mobility. New services are managed using old tools, which are not necessarily adequate to address the current market conditions or issues that may arise from smart mobility operations. For example, ridehailing platforms across the country are regulated as Private Hire Vehicles (PHVs) (the main difference from normal taxis is that PHVs can only be pre-booked and cannot be hailed on the street) and are licensed at the local authority level. The current system allows PHVs to operate anywhere in the country once they are registered with a local authority, making it difficult for authorities to control local standards. Similarly, local authorities can remove dockless bikes from the streets only using the Highways Act 1980, which gives them the power to remove unlawfully deposited bikes that cause nuisance or obstruction. However, this does not stop providers from launching their services in any location without any consultation with the local authority.

Collaboration appears to be a key response to the lack of effective regulation (see also Kronsell & Mukhtar-Landgren (2020), Chapter 7 in this book). At the moment, most agreements between providers and transport authorities are voluntary. For example, TfGM and Mobike, and TfWM and MaaS Global had signed Memorandums of Understanding setting out the principles of their relationships and the rules that the smart mobility providers in each case should follow, including alignment with local strategic priorities. However, these were agreements purely based on good will, without any consequences if they were not followed by either side. As such, all participants stressed the importance of maintaining good relationships with providers in order to avoid conflict such as that between Uber and TfL. Collaboration was also seen as a way to be 'kept in the loop' about the providers' decisions, including decisions to withdraw their services. Finally, most participants talked about using collaboration as a way to set the agenda for smart mobility providers. The participants from Milton Keynes Council (MKC) and TfWM said that they see themselves as enablers of services – not providers or operators – and that they encourage smart mobility providers to come and work together with them. Although they do not have direct control over smart mobility services, they use collaboration as a lever to attract services that are likely to have a positive impact on their cities. For example, transport authorities would choose to support providers by sharing their expertise on local transport and historic data but would withdraw this support if certain smart mobility services caused local disruption.

Nevertheless, uncertainty was a cross-cutting theme that emerged from the discussions about the current and future roles of the state in steering smart mobility. All participants expressed uncertainty about current conditions, for example, how the services that are already in place are being used and what is their impact on the transport network. They also expressed their concerns about how smart mobility may evolve in the future considering the apparent volatility of the market and also the rapid pace of technological development. Regarding the uncertainty about the current use and impacts, all participants highlighted the need to understand services better and quicker, although there were differences in the ways they approach this process of learning. Data sharing was an issue raised by all participants, but there seemed to be a lack of consensus over whether this should be voluntary or mandatory, and whether the data provided by smart mobility providers would be representative of different population groups. Most authorities focus their efforts on understanding the business case for scaling up different types of services through participating in trials, demonstrations and pilots. However, the focus is largely on understanding the potential advantages of services, and less on the potential issues they may create as part of the wider transport system. Aside from the issue of street clutter resulting from the operation of dockless micromobility, it was only the participant from TfGM who stressed the urgent need to understand the potentially negative impact of the operation of existing services, such as Uber's impact on congestion and public transport ridership, which may 'cannibalise the network'.

Uncertainty also seems to affect proactive action from the transport authorities. For example, some participants suggested that the quick changes in the micromobility market meant that they are hesitant to back a specific provider, mentioning that some discussions for potential collaboration had already fallen through after the providers withdrew from the British market. The participant from TfGM mentioned that they are turning down offers from providers who want to collaborate, as the organisation needs to take the time to understand the potential impacts of services within the local context. Some participants suggested that the uncertainty is too great to even guess the future role of the transport authority. For example, the participant from Leicester did not identify a future position as they considered very unclear how services such as MaaS could evolve for smaller cities like Leicester. The participant from WYCA also suggested that they take a 'wait and see' approach while they follow research and developments in the wider field of smart mobility and specifically around issues that would benefit the region, such as understanding how AVs and shared mobility services may affect the streetscape.

Future Accountability Regimes

Regarding the future accountability regimes and the role of each authority within them, the different participants envisaged a range of positions, as shown in Fig. 1. The key difference between these is *to whom* accountability is owed, with TfWM, MKC and WYCA leaning towards a hands-off approach, where the local authority does not intervene unless there is a problem, and TfGM considering essential that the local authority exerts more control over smart mobility services. The rationale behind a hands-off approach was that market competition is key and that ideally smart mobility providers should make profit, solve local problems and not cost money for the authority. There was a general agreement that under such an arrangement the transport authorities will continue to cover market failures through financial support for services operating in unprofitable areas or for specific user groups. In this case, smart mobility operators could be procured or tendered and would be accountable to the transport authority for meeting the terms of their contracts. On the other hand, the participant from TfGM noted that as an established authority that is accountable to and 'owned' by the public, it is necessary to have increased control in order to steer services and ensure better customer service from a system perspective.

In terms of the processes that would need to be followed to ensure accountability, the participants that envisaged a hands-off approach suggested that voluntary partnerships can be sufficient, based on evidence of existing successful partnerships with bus operators. They also suggested that when the authority is filling a gap in the market, existing processes such as subsidisation and tendering can work without necessarily the need of new policy tools. The participant from TfGM argued that new processes would need to be created at the local level, requiring further devolution of powers from the national to the local level. In addition, they suggested that accountability processes for smart mobility services would be shaped around the forthcoming reforms over bus services in the area, according to which the Combined Authority may take control of the deregulated market and become the franchising authority. At the time of writing, this decision had not been finalised.

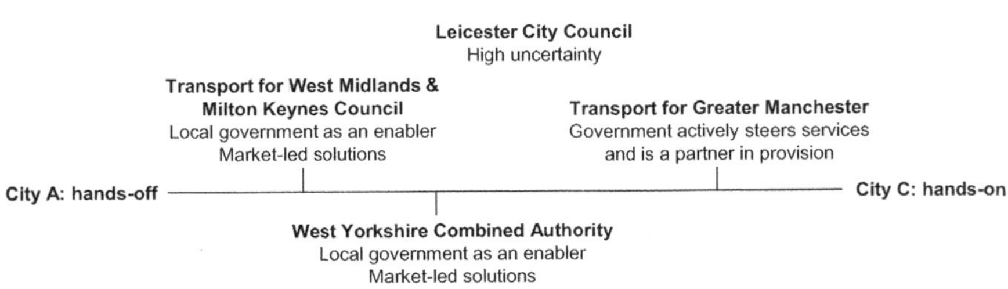

Fig. 1. Future Accountability Positions as Selected by the Interviewees.

It is not unexpected that the participants' views on accountability processes were high-level as there are few large-scale applications of smart mobility services in their areas and almost no proactive action to date to steer them. As such, the discussions around standards that the providers need to meet and the effects of non-compliance were also high-level and largely focussed on standards about passenger rights and safety, as well as some requirements on data sharing. Similarly to the present situation, future problems and the ways to address them were conceptualised mostly around specific services, such as clutter from dockless bikes, and less so at a system level. In addition, it was emphasised that in a competitive market environment there is less need for setting standards as users, through their choice, encourage the providers who do not perform well to improve their quality of service.

Regardless of the other elements of future accountability regimes, views on what purpose smart mobility should serve in the future were largely aligned among participants and with the narratives of providers. Participants spoke about services being new solutions to existing problems by providing the first and last mile of trips, integrated mobility ecosystems, seamless multimodality and more on-demand options for users. Most importantly, all participants believed that smart mobility services need to complement and to operate around the key mass transit network. In addition, particularly the participants who supported a hands-off approach assumed that services will align with local transport priorities and will nudge customers to make the 'right' choice. For example, one participant suggested that it would not be an issue for Uber to switch to an electric fleet or serve mobility hubs rather than compete with public transport, and that a MaaS operator would automatically discourage users from using car-based options as these would be the least profitable. In making these assumptions, the participants recognised that a greater level of control from the transport authorities will eventually be needed but discussed it as a distant need. The widespread adoption of AVs appeared to be trigger point for greater regulation, as it was considered that it will require central management of the road and kerb space, as well as limits imposed on models of ownership and sizes of fleets.

Accountability Barriers

The interviews demonstrated how the existing accountability arrangements act as a barrier for transport authorities' efforts to steer smart mobility in a way that maximises public value. English transport authorities outside London deal with a fragmented transport system, where there is limited integration between modes and little overall control over some services, such as buses. Therefore, smart mobility services need to be steered and plugged-in within an already messy context. In addition, the available policy tools are not always fit for purpose or sufficient to deal with the whole range of implications of smart mobility operations, leaving the authorities by default unable to steer services towards the direction they want (Moscholidou & Pangbourne, 2019, in press). An additional level of complexity is that smart mobility providers may have to adhere to varying processes and standards across the different local authorities, such as in the case of PHV licensing.

This complexity means that local authorities currently do not have overview, let alone control, of active smart mobility services. This automatically limits their ability to develop a detailed understanding of the impacts of services on the network and the users, and consequently constraints their ability to act proactively in steering providers.

The interviews also revealed that, to an extent, this complex status quo acts as barrier to proactive thinking and conceptualisation of how smart mobility could be managed as an integrated part of the transport system. Some participants' views of future conditions were bound by current limitations (e.g. 'dockless bikes can only be controlled through street regulations'), so they found it difficult to envisage alternative arrangements such as collaborative relationships with providers that do not solely rely on good will, or new policy tools, such as a permit system for services. Perhaps the reason why the participant from TfGM considered more powers over smart mobility essential for the future is because TfGM is exploring taking control of the local bus network, in which case they could be directly faced with the impact of competition from smart mobility services, such as Uber.

Of course, this is not simply an issue of imagination. Alternative arrangements are also practically difficult to implement due to intense budget constraints. All the participants who suggested a hands-off approach highlighted that a major advantage of this would be that any benefits of smart mobility operations would come at no cost to the local authority. Others mentioned that funding structures do not allow authorities to lead on the trials or research that would benefit them, as funders such as Innovate UK require projects to be led by academic teams. This means that authorities must rely on partners to also want to carry out these projects, which is often very challenging. Although not mentioned at the interviews, budget constraints have also affected the authorities' ability to attract and retain the needed numbers of skilled staff who would be able to drive the authorities' efforts to steer smart mobility. At the same time, the market is not mature enough to attract external funding from sponsors, making zero cost procurement very difficult for local authorities. For example, at the time of writing, TfWM had announced that they were terminating their partnership to launch a docked bikesharing system across the West Midlands as, among other reasons, their partner Nextbike had not secured the sponsorships needed under their agreement to launch the service (Madeley, 2019).

This complex regulatory and funding regime is further complicated by the fact that steering of smart mobility may seem unattractive to local politicians. Steering is often perceived as stifling innovation, which seems counterintuitive when local authorities in England want to signal that they are 'open for business'. Attracting innovation is widely seen as a success, and therefore local politicians do not want to deter providers, especially when there is little evidence on current or potential negative impacts of smart mobility on their cities. In addition, at a time when the narrative of smart mobility services is very dominant and presented as *the* future, both politicians and officials may take the view that since it is happening anyway, it is better to be part of it from the beginning in order to be better prepared for the time when it becomes the norm.

Conclusions: Need for a Rebalanced Narrative and Focussed Action

This chapter assessed the extent to which English transport authorities steer smart mobility services by analysing how they hold smart mobility providers accountable for the impacts of their services. The interviews revealed that the way local authorities see smart mobility services aligns more closely with the narrative of providers than the calls in academic literature to actively steer smart mobility. Although there seems to be an agreement that the state should steer smart mobility in a way that maximises public value, the participants suggested that, at least in the near future, steering can be done through collaborating with the providers in a competitive market context. When a more hands-on approach was considered appropriate, it was difficult for participants to identify the processes and standards that will achieve the alignment between smart mobility service provision and local transport goals. Uncertainty, the lack of evidence that the risks pointed out in the literature are valid, and limited adoption of different types of services certainly contribute to the participants' difficulty to fully identify the necessary accountability regimes for the future.

Although these interviews were only exploratory across a small sample of transport authorities, they already show that there is a need for multi-level action at the national, regional and local levels. It is also important to consider the implications of different local authorities wanting to take a significantly different stance in terms of regulation and control, whether this would bring further complexity in terms of governance, and what can be done at the national level to coordinate different approaches at the local level. Regardless of whether authorities take a more or less hands-on approach, there is a need for clarity about the regulatory role of the state and the role of the local authorities in managing different parts of the transition towards a low-carbon mobility future (Marsden & Docherty, 2019). As such, it is suggested that two key steps are necessary: rebalancing the narrative around smart mobility and focussed action.

The interviews revealed a sense of technological optimism and faith in the smart mobility market, and less focus on the potential risks that may come from the wider adoption of smart mobility services. Rebalancing the narrative around smart mobility can highlight the need for proactive steering. An important step towards this direction in the United Kingdom is the publication of the 'Future of Mobility: Urban Strategy'. The Strategy stresses the need for proactive action at the local level and highlights the social (safety, digital and financial exclusion), environmental (urban sprawl, increased congestion) and economic (abuse of monopoly power, market volatility) risks of failing to manage the change effectively within the short window of opportunity that is currently available (DfT, 2019). This rebalanced narrative also needs to be filtered down to regional and local governments and informs their strategic position, so that smart mobility services are proactively steered in ways that are adapted to the needs and priorities of cities and towns and help accelerate a wider low-carbon transition.

Moving forward, a rebalanced narrative needs to facilitate the development of clearer, focussed accountability regimes. In order to achieve this, it is necessary to overcome the specific barriers created by the current accountability arrangements for smart mobility services and the wider transport system. This research highlighted three key barriers for English local authorities: regulatory complexity, lack of integration in the existing system and lack of dedicated resources. At the moment, the DfT is undertaking a number of regulatory reviews focussing on different elements of transport innovation, including micromobility, MaaS, transport data and bus, taxi and PHV legislation. The regulatory reviews are expected to be a first step in untangling the existing complexity and facilitating integration across services provided by the private and the public sectors. In addition, the funding allocated for Future Mobility Zones will support further work on smart mobility applications.

Further action and future accountability arrangements need to reflect the rebalanced narrative and define the roles and responsibilities of providers and different levels of government through new, targeted regulations (*who is accountable and to whom*). It is recommended that to deliver public value, new accountability regimes need to allow cities to: (a) hold smart mobility providers accountable for any negative impacts through rules that are specifically related to different types of services (*effects*) and (b) monitor, assess and seize the benefits the innovation creates in order to achieve local transport goals and deliver a low-carbon transport system (*for what*). Within the new accountability regimes, cities can then develop accountability *processes and standards* that reflect the level and type of control they wish to have on providers.

References

Department for Transport (Dft). (2019). *Future of mobility: Urban strategy*.

Docherty, I., Marsden, G., & Anable, J. (2018). The governance of smart mobility. *Transportation Research Part A: Policy and Practice*, *115*, 114–125.

Dudley, G., Banister, D., & Schwanen, T. (2017). The rise of Uber and regulating the disruptive innovator. *The Political Quarterly*, *88*(3), 492–499.

Gunn, S. (2018). The history of transport systems in the UK. Retrieved from https://assets.publishing.service.gov.uk/government/uploads/system/uploads/attachment_data/file/761929/Historyoftransport.pdf

Kollewe, J., & McIntyre, N. (2019). Life cycle: Is it the end for Britain's dockless bike schemes? Retrieved from https://www.theguardian.com/cities/2019/feb/22/life-cycle-is-it-the-end-for-britains-dockless-bike-schemes

Kronsell, A., & Mukhtar-Landgren, D. (2020). Experimental governance of smart mobility. Some normative implications. In A. Paulsson & C. H. Sørensen (Eds.), *Shaping smart mobility futures – Governance and policy instruments in times of sustainability transitions* (pp. 119–135). Bingley: Emerald.

Legacy, C., Ashmore, D., Scheurer, J., Stone, J., & Curtis, C. (2019). Planning the driverless city. *Transport Reviews*, *39*(1), 84–102.

Lyons, G. (2018). Getting smart about urban mobility – Aligning the paradigms of smart and sustainable. *Transportation Research Part A: Policy and Practice*, *115*, 4–14.

Madeley, P. (2019). West Midlands 'Boris bike' scheme hanging in balance as nextbike ditched. Retrieved from https://www.expressandstar.com/news/transport/2019/07/25/boris-bike-scheme-hanging-in-the-balance-after-nextbike-is-ditched/

Marsden, G., & Docherty, I. (2019). Governance of UK transport infrastructures, Department for Transport. Retrieved from https://assets.publishing.service.gov.uk/government/uploads/system/uploads/attachment_data/file/780871/governance.pdf

Marsden, G., & Reardon, L. (Eds.). (2018). *Governance of the smart mobility transition*. Bingley: Emerald Publishing Limited.

Mashaw, J. L. (2006). Accountability and institutional design: Some thoughts on the grammar of governance. *Public Law Working Paper*, *116*, 115–156.

Moscholidou, I., & Pangbourne, K. (2019, in press). A preliminary assessment of regulatory efforts to steer smart mobility in London and Seattle. *Transport Policy*. https://doi.org/10.1016/j.tranpol.2019.10.015

Pangbourne, K., Mladenović, M. N., Stead, D., & Milakis, D. (2019). Questioning mobility as a service: Unanticipated implications for society and governance. *Transportation Research Part A: Policy and Practice*, in press. https://doi.org/10.1016/j.tra.2019.09.033

Reardon, L. (2020). Smart mobility as a catalyst for policy change towards low carbon mobility? In A. Paulsson & C. H. Sørensen (Eds.), *Shaping smart mobility futures – Governance and policy instruments in times of sustainability transitions* (pp. 139–151). Bingley: Emerald.

Sherman, L. (2019). Can uber ever deliver the transformative, profitable future that its CEO has promised? Retrieved from https://www.forbes.com/sites/lensherman/2019/06/05/can-uber-ever-deliver-the-transformative-profitable-future-that-its-ceo-has-promised/#16f3815c74af

Stone, J., Ashmore, D., Legacy, C., & Curtis, C. (2020). Challenges for government as facilitator and umpire of innovation in urban transport: The view from Australia. In A. Paulsson & C. H. Sørensen (Eds.), *Shaping smart mobility futures – Governance and policy instruments in times of sustainability transitions* (pp. 105–118). Bingley: Emerald.

Topham, G. (2018). Uber wins 15-month probationary licence to work in London. Retrieved from https://www.theguardian.com/technology/2018/jun/26/uber-case-licence-london

Torrance, D. (2019). Introduction to devolution in the UK. House of Commons Library Briefing paper. Retrieved from https://researchbriefings.parliament.uk/ResearchBriefing/Summary/CBP-8599#fullreport

Chapter 3

The Impacts of Automated Vehicles on the Transport System and How to Create Policies that Target Sustainable Development Goals

Anna Pernestål, Albin Engholm, Ida Kristoffersson and Johanna Jussila Hammes

ABSTRACT

Automated vehicles are likely to have significant impacts on the transport system such as increased road capacity, more productive/enjoyable time spent travelling in a car, and increased vehicle kilometres travelled. However, there is a great risk that automated driving may negatively impact the environment if adequate policies are not put in place. This chapter examines the effects of driverless vehicles and the types of policies required to attain sustainable implementation of the technology. To understand the effects on a systemic level, and to understand the needs and impacts of policies, the dynamics must be understood. Therefore, a causal loop diagram (CLD) is developed and analysed. One important insight is that the effects of driverless vehicles are mainly on the vehicular level (e.g., the reduced number of accidents per vehicle). These effects can be cancelled out on a systemic level (e.g., due to increased vehicle-kilometre travelled (VKT) that increases total number of accidents). The marginal costs of road transport are central to both freight and passenger transport. Automation will reduce marginal costs and shift the equilibrium in the transport system towards a state with higher VKT. This will lead to greater energy consumption and higher emissions. To attain sustainability goals, there might be a need to balance this

Shaping Smart Mobility Futures:
Governance and Policy Instruments in times of Sustainability Transitions, 37–53

doi:10.1108/978-1-83982-650-420201003

reduction of marginal costs by using policy instruments. In the work, CLDs is experienced to be a useful tool to support the collaboration between experts from different fields in the dialogue about policies.

Keywords: Automated vehicles; system dynamics; policy; causal loop diagram; collaboration tool; self-driving vehicles

Introduction

Automated vehicles are likely to have significant impacts on the transport system, and a growing body of research literature discusses these potential effects (such as increased road capacity, more productive/enjoyable time spent travelling in a car, and increased vehicle kilometres travelled). One conclusion from this research is that there is a great risk that driving automation may negatively impact the environment if adequate policies are not put in place.

To understand how automated vehicles should be governed to attain a transport system that fulfils the UN's Sustainable Development Goals (SDGs), it is important to realise that the effects listed in the literature are not isolated from nor independent of each other. Instead, they affect each other and are nestled in complex networks of causes and effects that often include time delays. This challenge is not unique to automated vehicles; Sterman (2001) gives several examples, including roadbuilding programmes designed to reduce congestion that have increased traffic, delays, sprawl, and pollution. To design policies that mitigate the possible negative effects of automated vehicles in the long term, these complex networks of causes and effects must be understood.

Complex networks and causes and effects can be modelled using system dynamics and causal loop diagrams (CLDs) to investigate the impacts of policies (Sterman, 2000). This chapter explains how CLDs can be used to describe the effects of introducing automated vehicles into a transport system, and how they can be used to analyse the needs for policies. A CLD that describes the impact of driving automation on a transport system is presented, including both freight and passenger transport. The CLD was created by a group of experts and is based on findings in the literature on the effects of driving automation. Needs and possibilities for governing an automated transport system are discussed based on this CLD.

Definitions and Methods

This section includes definitions of important concepts, a description of the literature searches performed to map existing knowledge about the effects of automated vehicles and policy needs, and an introduction to CLDs.

Definitions

Definition of driverless vehicles. The effects of automated vehicles are expected to be most significant when vehicles are fully driverless, either within a certain driving domain (SAE level 4[1]) or driverless on all roads (SAE level 5). Therefore, these higher automation levels have been chosen as the focus of this chapter, and these vehicles are referred to as driverless vehicles.

Definition of sustainability goals. In this chapter, the policy changes needed are structured and related to the SDGs. The SDGs have been rationalised by the Swedish Transport Administration (2018) into 10 transport-related goals, which are used in this chapter. Two of the goals deal with accessibility. The rest concern reliability, security, climate impact, biodiversity, air quality, noise, traffic safety, and active travel modes (walking and cycling).

Definitions of costs. Two related definitions of costs are used in this chapter. One is *generalised travel costs*. This refers to the monetised sum of all costs that an individual must bear in order to undertake their trip. For example, fuel costs, insurance, ticket costs, travel time, waiting time, etc., are included in generalised travel cost. The other is *marginal cost*. A marginal cost, which in mathematical terms is the first derivative of the generalised travel (or total) cost function, indicates the change in cost if one more unit is consumed. The marginal cost of one additional vehicle-kilometre travelled (VKT) would, on the vehicle level, be the cost of fuel and the wear and tear of the vehicle for that kilometre. The marginal cost of a VKT for society includes emission costs, increased congestion, increased accident cost, increased wear of the road, etc. The difference between private and social marginal cost gives rise to a need for policy instruments.

Literature Search

Several simulation studies have been published in recent years that investigate the effects of automated vehicles. The focus of these studies has mainly been on urban passenger transport in the form of (shared) driverless taxis, whereas the effects of automated vehicles in rural areas have been less investigated (but see Pettersson-Löfstedt & Khan, 2020, Chapter 11 in this volume). The effects described in this chapter are primarily based on two review articles, one by Soteropoulos, Berger, and Ciari (2019) and the other by Pernestål and Kristoffersson (2019). Both articles review a large number of simulations of driverless-vehicle case studies from all over the world and synthesise the reported effects. The two review articles are complemented with literature on the need for policies. Search terms 'policy instruments driverless vehicles', 'policy driverless vehicles', and 'policy autonomous vehicles' were used in the Summon database. Fifty-seven full-text articles were considered to be relevant out of several hundred articles found.

No comprehensive review study was found for automated freight transport. Therefore, a literature search was conducted in Scopus using the search terms 'Road Transport OR Transport OR Logistics AND Freight AND automated OR self-driving OR driverless OR autonomous' and 'truck AND automated OR self-driving OR driverless OR autonomous'. Snowballing was conducted from

[1]See SAE International (2018) for definitions of the different levels of automation.

identified relevant papers. Grey literature was included since the number of scientific publications on the subject is limited. Twenty-five articles were found.

Causal Loop Diagrams

A CLD is a map of key variables and causal relationships for the system under study. CLDs are typically used to bring structure to complex systems and to capture the interaction between several variables that describe different aspects (e.g., technical, economic, and behavioral) of the system, in contrast to focussing on one or a few relationships (Gruel & Stanford, 2016). CLDs consist of two types of components, variables, and causal links. CLDs, however, are not necessarily simple as they are used to describe complex systems. Fig. 1 shows an example of the components of a CLD. The arrows represent causal links between variables. The arrows originate from the causing variable and point to the affected variable. The arrows are indicated with a polarity sign (+ or −), which represents whether the causality is positive (i.e., an increase in the causing variable will increase the value of the affected variable) or negative (i.e., an increase in the causing variable will decrease the value of the affected variable). When two or more variables are linked together in a loop, they give rise to a feedback mechanism. Feedback loops can either be reinforcing (i.e., an increase in one variable will be amplified in the loop, which will yield even higher values) or balancing (i.e., the increase in one variable will be reduced in the loop). See, for example, Sterman (2000) for a more detailed description of CLDs.

System dynamics originated as a tool for business management research but has emerged as a modelling approach in transport research with two of the most common application areas being the analysis of the uptake of alternate fuel vehicles and strategic policy analysis (Shepherd, 2014). More recent research has utilised system dynamics to study the market diffusion (Nieuwenhuijsen, Correia, Milakis, van Arem, & van Daalen, 2018) and system impacts (Pfaffenbichler, Gühnemann, Klementschitz, Emberger, & Shepherd, 2019) of driverless vehicles.

A CLD that describes the impacts of the automation of passenger and freight transport on the national level for Sweden is presented in this chapter. The CLD was developed using group model building (Andersen, Richardson, & Vennix, 1997) during a full-day workshop that engaged a focus group consisting of the

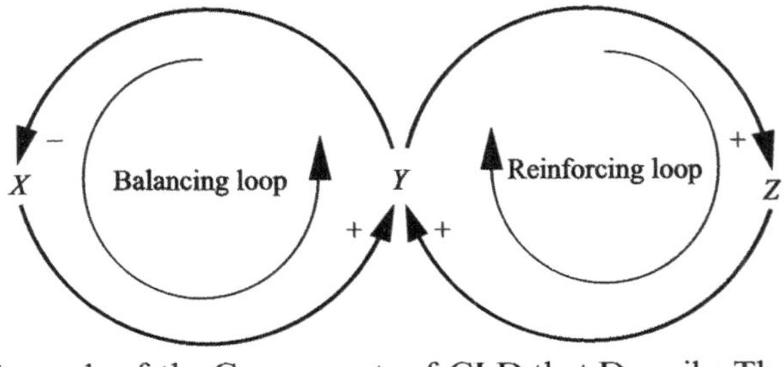

Fig. 1. Example of the Components of CLD that Describe Three Related Variables, *X*, *Y*, and *Z*.

authors and representatives from the Swedish Transport Administration. The focus group participants are experts in technology development, driverless vehicles, economics, transport modelling, and transport sustainability analysis. The purpose of the workshop was to identify and analyse the most important dynamic effects on the transport system that result from the introduction of driverless vehicles. A small focus group with extensive expertise in different core areas was preferred over a larger group. The workshop was performed using a set of standardised group model building scripts (Scriptapedia, 2019). The literature studies presented earlier in the chapter provided the knowledge base for the development of the CLD. By definition, the CLD resulting from the workshop was influenced by the expertise and the mental models of the focus group participants, and a focus group with other participants would have generated a different CLD.

Background Literature on Impacts and Needs for Policy Instruments

What does the Literature say about the Effects of Automated Vehicles?

Passenger Transport. The two recent literature reviews – Pernestål and Kristoffersson (2019) and Soteropoulos et al. (2019) – which summarise the simulated effects of driverless vehicles for passenger transport, basically arrive at similar conclusions: First and foremost, driverless vehicles are found to increase both person-kilometres travelled (PKT) and VKT, especially if rides are not shared. The increase in VKT is mainly due to the relocation of or the cruising done by empty vehicles, new user groups such as the elderly and young people gaining access to cars, and increased demand for more and longer trips. Increased demand is caused by car travel becoming more attractive as an expected reduction in the value of travel time (VTT) decreases both the generalised and the marginal cost for travelling by a car, see also Wadud, MacKenzie, and Leiby (2016) and Fraederich, Heinrichs, Bahamonde-Birke, and Cyganski (2019). Related to this is the effect of urban sprawl, namely that commuters are willing to accept longer travel times and, thus, may relocate live further away from an urban centre. Second, driverless vehicles are found to reduce the demand for walking, biking, and travelling by public transport. On the other hand, driverless buses are found to lower operation costs for public transport. As a force that counteracts urban sprawl, an increase in the efficiency of public transport has the potential to speed up urbanisation. Third, driverless taxis are found to have the potential to reduce the total number of vehicles needed for passenger transport. However, it should be noted that (shared) driverless taxi applications in urban areas have been well studied in simulations, whereas driverless long-distance travel, driverless rural applications, and driverless public transport applications have been much less studied, and the effects of these applications are less understood.

Furthermore, driverless vehicles are expected to have an impact on the demand for and localisation of parking. The vehicles are expected to drive themselves outside of city centers for parking, given that the cost of providing parking infrastructure in city centers exceeds the cost of the trip to a free parking place outside the centre (Chapin, 2016; Fraederich et al., 2019; Zakharenko, 2016). The development of driverless taxi services is expected to reduce the demand for parking in city centers (Heinrichs, 2016), while demand for drop-off and pick-up zones will increase (Marsden, Docherty, & Dowling, 2019). While this demand has positive effects on, for example, increased livability in city centers when former parking spots can be converted to broader and better walkways and bicycle lanes and green areas (Alessandrini, Campagna, Delle Site, Filippi, & Persia, 2015; Chapin, 2016; Faisal, Yigitcanlar, Kamruzzaman, & Currie, 2019), it also has negative consequences for the income of cities, and possibly also for the emissions of climate gases, air pollution, noise, and congestion.

Driverless vehicles are also expected to contribute to more energy efficient driving (Faisal et al., 2019; Mersky & Samaras, 2016), reduced congestion due to improved traffic flow (Wadud et al., 2016), homogenisation of traffic flow (Bagloee, Tavana, Asadi, & Oliver, 2016; Wadud et al., 2016), reduced sudden use of brakes (Fagnant & Kockelman, 2015), and right-sizing of the vehicle choice (Wadud et al., 2016). At the same time, many of these effects are uncertain.

Freight Transport. Automated vehicles in freight transport are most likely to first be used for long-haulage transport on highways and for industrial transport flows (Flämig, 2016; International Transport Forum, 2017; Kristoffersson & Pernestål Brenden, 2018; Roland Berger, 2016; Viscelli, 2018). The reasons for this are primarily that these traffic environments are less complex than urban environments, that these flows have high volumes of goods, and that they are repetitive. In addition, there are often only a few actors involved, which simplifies the introduction of new business models.

For several reasons, it is likely that the breakthrough for driverless vehicles will come earlier for freight transport, and that uptake will be faster for freight transport than for passenger transport. First, in contrast to people's private transport choices, freight transport decisions are primarily based on rational and cost-minimising principles, and removing the driver would have a substantial impact on transport cost (Kristoffersson & Pernestål Brenden, 2018; Lunkeit, Flämig, & Rosenberger, 2019; Müller & Voigtländer, 2019). Second, the expected future lack of truck drivers is an important factor that increases the speed of transition. Third, obstacles such as safety, acceptance, and trust are less significant for transporting goods than for transporting people.

The literature that analyses the effects of driverless freight transport remains limited, but one conclusion that can be made is that operational costs for road freight transport with heavy duty trucks can be reduced by around 30% (Engholm, Pernestål, & Kristoffersson, 2019). Additional savings can be achieved by platooning. Typical estimates of fuel savings for trucks during platooning are 4–15% (Flämig, 2016; Kückelhaus, 2014; Manyika et al., 2013; Muratori et al., 2017). These cost savings will lead to an increase in the demand for road freight transport and likely also to a shift from rail and maritime transport to road transport. For example, Bao and Mundy (2018) estimate that 25–40% of US rail freight transport could shift to road if driverless vehicles are implemented.

Summary of needs for Policy Instruments

Once the effects of driverless vehicles have been identified, policy instruments to mitigate negative externalities can be designed. As noted above, the goals set by the Swedish Transport Administration (2018) are used as a starting point for policy assessment.[2] Needs for policy instruments are summarised in Table 1.

The positive aspects in Table 1 arise if driverless vehicles are introduced into the transport system, and if this is the case, no new policies will be needed to achieve the gains. A need for policy instruments arises if the private marginal cost, that is, the marginal cost of an individual using a driverless vehicle, is lower

Table 1. Transport Sector Goals, How They are Impacted by Driverless Vehicles, and the Need for Policy Instruments to Internalise External Effects.

Goal	Impact from Driverless Vehicle (±)	Possible Needs for Policy Instruments
Accessibility in an entire country[a]	+ Lower transport costs	* Increased subventions to maritime and rail transport
	+ Less driver shortage	* Oversight of employment conditions of transport workers
	− Reduction of the comparative advantage of maritime and rail transport	* Potential regulation of market power (monopoly/oligopoly) on the automated freight transport market[b]
	− Increased driver unemployment	
Accessibility for all	+ Increased mobility for people with disabilities and for citizens regardless of age	* Market design for shared driverless vehicle fleets, potential regulation of market power (monopoly/oligopoly) on the autonomous taxi market[c]
	− Reduced demand for public transport	* Policy instruments to reduce VKT, depending on the benefits and costs of increased motorised travel
	− Increased VKT, and thus increased congestion and other externalities	

[2]In Swedish Transport Administration (2018), there are 10 goals, but the impacts of the goals air quality and noise are expected to be similar for driverless vehicles and manually driven vehicles. Therefore, these two goals are not analysed further.

Table 1. (*Continued*)

Goal	Impact from Driverless Vehicle (±)	Possible Needs for Policy Instruments
Reliability and ease of use	+ Reliability and comfort increase, trip planning becomes easier	* Cybersecurity must be ascertained
	+ Reduced vulnerability to disturbances	* Policy instruments/policies for trip planning and payment solutions
	− The entire transport system becomes more susceptible to disturbances	
Security	+ Door-to-door transport can increase security	* Regulation to increase security in shared vehicles
	− Shared vehicles can be (experienced as) unsafe	
Impact on the climate	− Due to increased VKT, but depending on the fuel used	* If electric vehicles: no new policy instruments required besides relevant climate policies[c]
	± Impacts from the production of vehicles and batteries, depending on the impact of total fleet size	* If fossil fuels, marginal pricing so that the marginal climate damage equals the marginal cost of emissions abatement
		* Reduced need for new infrastructure, at least in the short term
Biodiversity	+ Reduction in dead animals in traffic as driverless vehicles can avoid collisions	–
Traffic safety	+ Increases	–
Active travel modes	− Motorised transport modes increase, and non-motorised modes decrease	* Increase the marginal cost of motorised transport modes * Regular measures to increase walking and cycling

[a]One of the sub-goals of accessibility is that the transport industry has fair employment practices and healthy competition within modes.

[b]It is not illegal for a firm to have market power; it is the abuse of market power that is illegal according to the Swedish Competition Act and Article 102 of the Treaty on the Functioning of the European Union. Abuse is defined as taking advantage of market power in order to distort competition and, in the end, reduces consumer surplus, which creates a deadweight loss.

[c]Relevant climate policies refer to the existence of, for example, an emissions trading system, or CO_2 tax, that internalises the climate externality arising from electricity production.

Source: The table is an updated version from Jussila Hammes (2019).

than the societal marginal cost, or if there are some other market imperfections, such as market power. The policy instruments in the right-most column aim to ascertain whether or not the introduction of driverless vehicles is positive, not only for individuals but also for an entire society. They do not consider the question of whether or not the market, alone, introduces driverless technology.

A CLD for Driverless Assenger and Freight Transport

Overview of the CLD

A CLD was created to capture the effects of the introduction of driverless vehicles on both passenger and freight transport by describing the most important dynamics of the system without the introduction of any new policies, see Fig. 2. The CLD is composed of three sub-models: (1) a passenger transport sub-model intended to capture effects, such as mode shifts, car access, and transport demand; (2) a freight transport sub-model that captures mode shifts between rail and road transport, fill rates for trucks, and freight transport demand; and (3) the shared transport system sub-model where the passenger and freight sub-models are connected, consisting of road infrastructure and other resources shared and influenced by the passenger and freight transport systems. The shared transport system sub-model captures the effects of shared mobility on traffic volumes, congestion, the share of electric vehicles, emissions, and traffic safety. In Fig 2, variables that can be directly linked to the transport sector goals in Table 1 are marked with boxes.

The CLD in Fig. 2 is complex and contains many variables and dynamics, so a comprehensive summary is not possible. Instead, an analysis of the causal relationships related to the variable *Total passenger transport VKT* is presented as an example of the dynamics of the system. This variable is important for the system since it interacts with many other variables, and it is a key metric for the transport system.

Total passenger transport VKT is caused by three variables of which *Total PKM (Passenger km) demand passenger transport* is one. *Total PKM demand passenger transport*, in turn, is determined by the level of *Urban sprawl* and the *Average marginal cost passenger transport. Total passenger transport VKT* directly impacts *Congestion,* which, in turn, impacts the *Average marginal cost passenger transport. Congestion* is also impacted indirectly by *Total passenger transport VKT* via the impact on *# traffic accidents.*

The above causes and effects related to *Total passenger transport VKT* show that there are two feedback mechanisms in the system that have a balancing effect on VKT, as follows:

(1) *Total passenger transport VKT* → + *Congestion* → + *Average marginal cost passenger transport*→ – *Total PKM demand passenger transport* → + *Total passenger transport VKT*
and
(2) *Total passenger transport VKT* → + *# traffic accidents* → + *Congestion* → + *Average marginal cost passenger transport* → – *Total PKM demand passenger transport* → + *Total passenger transport VKT*

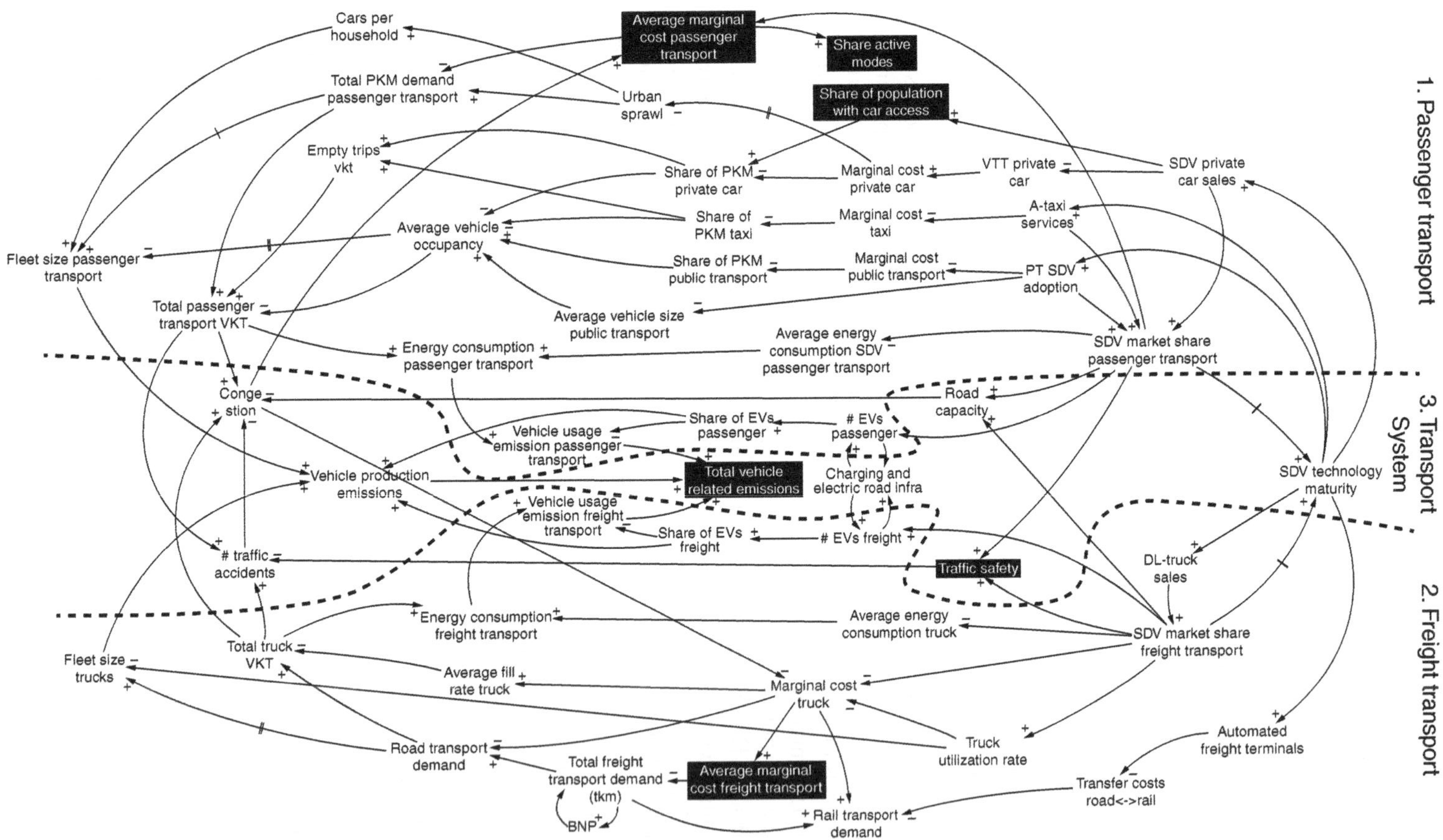

Fig. 2. A CLD that Captures the Effects and Dynamics of the Introduction of Driverless Vehicles into a Transport System.

Analysis of the CLD

To enable a quantitative analysis of the effects of driverless vehicles, the CLD must be transformed into a stock-flow diagram (Sterman, 2000), where each causal relation is described with numerical equations so that a formal model of the system can be obtained and simulated. However, valuable qualitative insights can be gained from the CLD in Fig 2:

(1) The marginal cost of cars and road freight vehicles are central variables in several of the feedback loops in the CLD. Driving automation will shift the demand–supply equilibrium in the transport system towards a state with higher levels of VKT. This is in line with findings in previous simulation-based research.
(2) The majority of research findings agree that VTT will decrease with driverless vehicles, however, to what extent is more uncertain (Correia, Looff, Cranenburgh, van Snelder, & van Arem, 2019; Kolarova, Steck, Cyganski, & Tommer, 2018; Steck, Kolarova, Bahamonde-Birke, Trommer, & Lenz, 2018). Nevertheless, it is important to consider this uncertainty in the design of policies for driverless vehicles. It is also important to collect more information about VTT as technology is developed and more services become available.
(3) VTT will have a major impact on the marginal cost of passenger transport and, thus, the expected decrease in VTT resulting from driving automation will lead to a fall in the marginal cost of road transport. This results in longer travel times (longer routes and more in-vehicle time) and more trips done by car. The CLD shows that this will lead to more VKT and, with a delay, possibly to urban sprawl, which further increases VKT.
(4) Congestion in the CLD is the only factor that balances the increase in VKT induced by driving automation. Thus, other policies to reduce increased VKT may be needed, for example, policies to increase the marginal cost of driving in order to curb the negative externalities caused by VKT.
(5) Driverless vehicles are expected to have several positive effects on the transport system, for example, they are safer and can improve traffic flow. However, these effects are typically on the vehicle level (i.e., fewer number of accidents per vehicle) or are effects that further increase the attractiveness of road transport (e.g., improved traffic flow). On a system level, these positive effects may be offset by systemic interactions (e.g., due to the increase in VKT). These system effects are complex as some effects (e.g., land use and relocation) have long time delays.
(6) There are several mechanisms that will influence vehicle fleet size as a result of the introduction of driverless vehicles. The expected increase in transport demand resulting from driving automation because of lower marginal cost will require a larger vehicle fleet, everything else being equal. The greater demand for freight transport will be offset by an increase in truck productivity for driverless trucks, meaning that fewer trucks will be required to perform a given freight transportation task.

(7) Similarly, simulation studies show that fleets of shared driverless vehicles could drastically reduce the required fleet size for a given demand for passenger transport compared to conventional cars. However, several challenges remain to be overcome in order for shared vehicles to gain ground, for example, the booking system must be easy to use, vehicles must be available on-demand, the vehicles must be clean, and it will not be possible to store personal belongings in the vehicle, and people must trust the shared driverless vehicle service. These challenges must be addressed without increasing the cost of the service too much in comparison with conventional cars. It is also possible that the urban sprawl induced by passenger travel will lead to increased car dependency, which will result in more cars per household and, thus, an increase in fleet size.

(8) Since the electricity mix in Sweden is currently low in carbon emissions compared to many other nations, electric vehicles reduce greenhouse gas emissions from road transport substantially (i.e., emissions from VKT). However, the emissions during the production phase of electric vehicles are substantial. Thus, climate emissions from the lifecycle perspective of a road transport vehicle will be more important to consider in a future with electric vehicles. The number of produced vehicles and the share of electric vehicles are necessary variables to understand total lifecycle vehicle-related emissions.

Discussion

Reflections on Needs for Policies

As driverless vehicles reduce the marginal cost of driving, make car travel possible for new user groups, and make freight transport cheaper, the accessibility to sites, activities, and products will increase for many people and businesses, which is in line with the two accessibility goals set by the Swedish Transport Administration (see Table 1). However, this leads to increased levels of VKT both for passenger and freight transport, which would have a negative impact on several of the other goals, including climate, noise, and travel by active modes. This is an example of conflicting goals, and increased accessibility and reduced externalities must be balanced when considering policy implementation.

Another concern is policies for parking. As noted above, driverless vehicles are expected to reduce demand for parking in city centers. At the same time, the number of drop-off and pick-up zones is expected to increase. Therefore, future research should consider the optimal land use for parking in and outside of cities, drop-off and pick-up zones, and the optimal pricing for all three. Such research should at least weigh in the social cost of parking and drop-off/pick-up space in and outside of city centers, increased externalities from empty trips to a parking spot, reduced cruising for parking (Shoup, 2005), the consequences for municipal income and taxes, and the deadweight loss of taxes.

One question is whether policy instruments should be used to speed up the uptake of driverless technology. Economic theory stipulates that some type of market imperfection must be present to justify a society's support for a technology.

Driverless technology seems, to a large extent, to be driven by private car manufacturers and by demand from both private individuals, and above all, the freight transport industry. While driverless technology has societal benefits, several of the benefits mainly replace existing means of achieving the same result, for example, subsidised trips for persons with disabilities and the elderly. Nevertheless, it is possible that the development of infrastructure (e.g., 5G technology) will have positive impacts that the private sector does not take into consideration. If this is so, some form of public support for the building of a network could be justified.

Reflections from a Sustainability Perspective

As noted above, several authors expect driverless vehicles to lead to a reduction in the negative environmental impacts that arise from road transport. These impacts are uncertain, however. Faisal et al. (2019) note that energy consumption and emissions from driverless vehicles do not follow directly from automatisation but arise because of reconstruction of the driving cycle, vehicle design, choice of propulsion technology, policies, and the design of the transport systems. The impacts are, therefore, a secondary and not a primary one.

There is also a risk for increased negative externalities with the introduction of driverless vehicles. The VKT for both freight and passenger transport is expected to increase. This would, in turn, increase traffic volumes and congestion. Greenhouse gas and air pollution emissions increase with increased VKT, either directly from fossil fuels burned in a combustion engine or if electricity is produced from fossil fuels. The manufacture of batteries for electric vehicles is also a source of emissions. Onboard electronic systems, both for driving and for entertainment, may additionally increase energy consumption (Gawron, Keoleian, De Kleine, Wallington, & Kim, 2018).

Several authors argue that there is a natural synergy between the fleets of driverless and electric vehicles (Bagloee et al., 2016; Chen & Kockelman, 2016; Wadud et al., 2016) as driverless vehicles have the potential to solve some of the limitations of electric vehicles, for example, limited range, access to charging infrastructure, and charging time. Fleets of driverless vehicles would reduce the disutility created by these factors by controlling driving distance and charging activities based on real-time demand and available charging infrastructure. These benefits apply primarily to fleets of shared vehicles, however, and are of less importance to privately owned vehicles. Furthermore, electrification may lower the barriers for new vehicle manufacturers to enter the market and shift competition from the development of combustion engines to other technologies, including automation.

Governance and policies are needed to ensure that automation contributes to a sustainable transport system. Based on the literature review and the CLD developed in this chapter, measures could, above all, include policies to balance the decrease in marginal costs to avoid a too large increase in VKT, especially in congested areas and during peak hours, for example, a km-based road tax that varies according to time, place and emission levels from the vehicle. However, before such policies can be implemented, a large number of questions must be resolved relating to the cost-efficiency of tax design, impact on behaviour, interactions with other (existing) policies, tax acceptability to the electorate, the political economics of such a tax, issues of integrity, etc. One way to perform analyses might be with the help of CLDs and simulations using stock-flow diagrams.

CLD as a Collaboration Tool

As discussed by McLeod, Curtis, and Stone (2020, Chapter 5 in this volume) tools where other aspects than those in traditional transport models are needed to develop policies for smart mobility. One key benefit of the CLD methodology is that it encourages and facilitates cooperation among experts from different fields. Building a CLD promotes the development of a shared mental model of the dynamics of the system under study among the participants. This improves the potential to precisely discuss the behaviour of the system and reflect on the potential impacts of changes to the system, for example, through interventions. With the CLD the discussion is held on a high level and, thus, does not require the participants to understand the specific methods and models of a specific field. Neither do the participants need to learn the specific terminology for each area, since the CLD formalises the effects in terms of polarity and feedback loops. For example, the field of transport analysis uses advanced models and specific terminology that describe how trip generation depends on generalised travel costs (including travel time weighted by VTT), income, etc., while shifts in the demand and supply of VKTs can easily be understood and described within the framework of economic theory. However, it is not necessary to understand the details of either model to create a CLD. CLD methodology is not only beneficial to the participants developing it; it is also easier to communicate a CLD model to politicians and stakeholders than conventional transport and economic models. Thus, the methodology has pedagogic advantages.

Conclusions

This chapter examined the effects of driverless vehicles and the types of policies required to attain sustainable implementation of the technology. A literature review and an analysis of needs for policy instruments were performed. Research literature describes and discusses current positive and negative effects of automation, but to understand the effects on a systemic level and to understand the needs and impacts of policies, the dynamics must be understood. Therefore, a CLD was developed and analysed.

Main insights from the analysis include the following aspects:

- The positive effects of driverless vehicles are mostly related to one single vehicle (e.g., the reduced number of accidents per vehicle; increased capacity, as a single vehicle requires less road space; and increased accessibility, as new user groups gain access to cars). These effects can be cancelled out on a systemic level (e.g., due to increased VKT and congestion that decreases accessibility).
- The marginal costs of road transport are central to both freight and passenger transport. Automation will reduce marginal costs and shift the equilibrium in the

transport system towards a state with higher VKT. This will lead to greater energy consumption and higher emissions. To attain sustainability goals, there might be a need to balance this reduction of marginal costs by using policy instruments.

- Electrification and other fossil-free propulsion can, to some extent, decouple VKT from emissions. Exploring policies that draw on correlations between electrification and automation would be interesting. At the same time, not only does VKT have an impact on climate, but so does the production of new vehicles and batteries. Therefore, fleet size, or more precisely, the number of produced vehicles should be taken into account when considering climate impacts.
- The analysis of the introduction of driverless vehicles highlights the conflicts between different sustainability goals, for example, accessibility versus climate, and these conflicts must be handled.

Designing and implementing policy instruments for new technologies, such as driverless vehicles, is a challenging task. To succeed, competence and knowledge from various disciplines must be integrated. This chapter showed how CLDs is a powerful tool to foster such collaboration and to reduce communication barriers between experts from various fields.

References

Alessandrini, A., Campagna, A., Delle Site, P., Filippi, F., & Persia, L. (2015). Automated vehicles and the rethinking of mobility and cities. *Transportation Research Procedia*, *5*, 145–160.

Andersen, D. F., Richardson, G. P., & Vennix, J. A. M. (1997). Group model building: Adding more science to the craft. *System Dynamics Review*, *13*(2), 187–201. https://doi.org/10.1002/(SICI)1099-1727(199722)13:2<187::AID-SDR124>3.0.CO;2-O

Bagloee, S. A., Tavana, M., Asadi, M., & Oliver, T. (2016). Autonomous vehicles: Challenges, opportunities, and future implications for transportation policies. *Journal of Modern Transportation*, *24*(4), 284–303.

Bao, K., & Mundy, R. A. (2018). *Emerging freight truck technologies: Effects on relative freight costs final report* (p. 30). Ames, IA: Institute for Transportation Iowa State University.

Chapin, T. (2016). *Envisioning Florida's future: Transportation and land use in an automated vehicle world*. Tallahassee, FL: Florida Department of Transportation.

Chen, T. D., & Kockelman, K. M. (2016). Management of a shared autonomous electric vehicle fleet: Implications of pricing schemes. *Transportation Research Record: Journal of the Transportation Research Board*, 2572, 37–46.

Correia, G., Looff, E., Cranenburgh, S. van Snelder, M., & van Arem, B. (2019). On the impact of vehicle automation on the value of travel time while performing work and leisure activities in a car: Theoretical insights and results from a stated preference survey. *Transportation Research Part A: Policy and Practice*, *119*, 359–382. https://doi.org/10.1016/j.tra.2018.11.016

Engholm, A., Pernestål, A., & Kristoffersson, I. (2019). *Cost analysis of driverless truck operations*.

Fagnant, D. J., & Kockelman, K. M. (2015). Preparing a nation for autonomous vehicles: Opportunities, barriers and policy recommendations. *Transportation Research Part A: Policy and Practice*, *77*, 167–181.

Faisal, A., Yigitcanlar, T., Kamruzzaman, M., & Currie, G. (2019). Understanding autonomous vehicles: A systematic literature review on capability, impact, planning and policy. *The Journal of Transport and Land Use*, *12*(1), 45–72.

Flämig, H. (2016). Autonomous vehicles and autonomous driving in freight transportation. In *Autonomous Driving*.

Fraederich, E., Heinrichs, D., Bahamonde-Birke, F. J., & Cyganski, R. (2019). Autonomous driving, the built environment and policy implications. *Transportation Research Part A*, *122*, 162–172.

Gawron, J. H., Keoleian, G. A., De Kleine, R. D., Wallington, T. J., & Kim, H. C. (2018). Life cycle assessment of connected and automated vehicles: Sensing and computing subsystem and vehicle level effects. *Environmental Science & Technology*, *52*(5), 3249–3256.

Gruel, W., & Stanford, J. M. (2016). Assessing the long-term effects of autonomous vehicles: A speculative approach. *Transportation Research Procedia*, *13*, 18–29.

Heinrichs, D. (2016). Autonomous driving and urban land use. In M. Maurer, J. C. Gerdes, B. Lenz, & H. Winner (Eds.), *Autonomous driving. Technical, legal and social aspects* (pp. 213–231). Heidelberg: Springer Open.

International Transport Forum. (2017). Managing the transition to driverless road freight transport [case-specific policy analysis]. Retrieved from https://www.itf-oecd.org/sites/default/files/docs/managing-transition-driverless-road-freight-transport.pdf

Jussila Hammes, J. (2019). *Styrmedel för självkörande fordon* [*Policy instruments for autonomous vehicles*] (No. S-WoPEc 2019:3). Stockholm.

Kolarova, V., Steck, F., Cyganski, R., & Tommer, S. (2018). Estimation of the value of time for automated driving using revealed and stated preference methods. *Transportation Research Part A: Policy and Practice*, *31*, 35–46. https://doi.org/10.1016/j.tra.2018.11.016

Kristoffersson, I., & Pernestål Brenden, A. (2018). Scenarios for the development of self-driving vehicles in freight transport. In *Proceedings of 7th transport research arena TRA 2018*, April 16–19, Vienna, Austria. Accepted for Presentation at the Transport Research Arena.

Kückelhaus, D. M. (2014). *Self-driving vehicles in logistics*. DHL Customer Solutions & Innovation 53844. Troisdorf: Germany.

Lunkeit, S., Flämig, H., & Rosenberger, K. (2019). The impact of driverless vehicles on transportation costs in road haulage: A system dynamical approach. In U. Clausen, S. Langkau, & F. Kreuz (Eds.), *Advances in production, logistics and traffic* (pp. 116–127). Cham: Springer. https://doi.org/10.1007/978-3-030-13535-5_9

Manyika, J., Chui, M., Bughin, J., Dobbs, R., Bisson, P., & Marrs, A. (2013). *Disruptive technologies Advances that will transform life, business, and the global economy*. New York, NY: MGI, McKinsey & Company.

Marsden, G., Docherty, I., & Dowling, R. (2019). Making space for new mobility services? The curb as a critical boundary object. *Land Use Policy*, in press.

McLeod, S., Curtis, C., & Stone, J. (2020). Planning urban futures for autonomous and shared vehicles: The role of planning support tools as a policy instrument. In A. Paulsson & C. H. Sørensen (Eds.), *Shaping smart mobility futures – Governance and policy instruments in times of sustainability transitions*. (pp. 75–103) Bingley: Emerald.

Mersky, A. C., & Samaras, C. (2016). Fuel economy testing of autonomous vechiles. *Transportation Research Part C*, *65*, 31–48.

Müller, S., & Voigtländer, F. (2019). Automated trucks in road freight logistics: The user perspective. In U. Clausen, S. Langkau, & F. Kreuz (Eds.), *Advances in production, logistics and traffic* (pp. 102–115). Cham: Springer. https://doi.org/10.1007/978-3-030-13535-5_8

Muratori, M., Holden, J., Lammert, M., Duran, A., Young, S., & Gonder, J. (2017). Potentials for platooning in U.S. highway freight transport. *SAE International Journal of Commercial Vehicles*, *10*(1), 45–49. https://doi.org/10.4271/2017-01-0086

Nieuwenhuijsen, J., Correia, G. H. de A., Milakis, D., van Arem, B., & van Daalen, E. (2018). Towards a quantitative method to analyze the long-term innovation diffusion of automated vehicles technology using system dynamics. *Transportation Research Part C: Emerging Technologies*, *86*, 300–327. https://doi.org/10.1016/j.trc.2017.11.016

Pernestål, A., & Kristoffersson, I. (2019). Effects of driverless vehicles – Comparing simulations to get a broader picture. *European Journal of Transport and Infrastructure Research (EJTIR)*, *19*(1), 1–23. Retrieved from https://d1rkab7tlqy5f1.cloudfront.net/TBM/Over%20faculteit/Afdelingen/Engineering%20Systems%20and%20Services/EJTIR/Back%20issues/19.1/362018%20Pernestal.pdf

Pettersson-Löfstedt, F., & Khan, J. (2020). Smart public transport in rural areas: prospects, challenges and policy needs. In A. Paulsson & C. H. Sørensen (Eds.), *Shaping smart mobility futures – Governance and policy instruments in times of sustainability transitions* (pp. 187–201). Bingley: Emerald.

Pfaffenbichler, P., Gühnemann, A., Klementschitz, R., Emberger, G., & Shepherd, S. (2019). Societal impacts of highly automated cars using the toolbox of System Dynamics. Presented at the European Transport Conference, October 9. Retrieved from https://aetransport.org/past-etc-papers/search-all-etc-conference-papers?abstractId=6334&state=b

Roland Berger. (2016). *Automated trucks – The next big disruptor in the automotive industry?*, Chicago, IL: Roland Berger.

SAE International. (2018). Taxonomy and definitions for terms related to driving automation systems for on-road motor vehicles. Retrieved from https://webstore.ansi.org/Standards/SAE/SAE30162018?source=blog. Accessed on 25 June 2019.

Scriptapedia. (2019). Scriptapedia. Retrieved from https://en.wikibooks.org/wiki/Scriptapedia

Shepherd, S. P. (2014). A review of system dynamics models applied in transportation. *Transportmetrica B: Transport Dynamics*, *2*(2), 83–105.

Shoup, D. (2005). *High cost of free parking*. Chicago, IL: Planners Press.

Soteropoulos, A., Berger, M., & Ciari, F. (2019). Impacts of automated vehicles on travel behaviour and land use: An international review of modelling studies. *Transport Reviews*, *39*(1), 29–49.

Steck, F., Kolarova, V., Bahamonde-Birke, F., Trommer, S., & Lenz, B. (2018). How autonomous driving may affect the value of travel time savings for commuting. *Transportation Research Record: Journal of the Transportation Research Board*, *2672*(46), 11–20.

Sterman, J. (2001). System dynamics modeling: Tools for learning in a complex world. *IEEE Engineering Management Review*, *43*(4), 8–25.

Sterman, J. D. (2000). *Business dynamics—Systems thinking and modeling for a complex world*. Boston, MA: McGraw-Hill Higher Education.

Swedish Transport Administration. (2018). *Tillgänglighet i ett hållbart samhälle—Målbild 2030* [Accessibility in a sustainable society - Goals 2030] (No. 2018:235). Borlänge: Trafikverket.

Viscelli, S. (2018). *Driverless? Autonomous trucks and the future of the American trucker* (p. 73). Berkely, CA: Center for Labor Research and Education, University of California, Berkely and Working Partnerships USA.

Wadud, Z., MacKenzie, D., & Leiby, P. (2016). Help or hindrance? The travel, energy and carbon impacts of highly automated vehicles. *Transportation Research Part A: Policy and Practice*, *86*, 1–18.

Zakharenko, R. (2016). Self-driving cars will change cities. *Regional Science and Urban Economics*, *61*, 26–37.

Part II

Chapter 4

Crafting Effective Policy Instruments for 'Smart Mobility': Can Multi-level Governance Deliver?

Iain Docherty

ABSTRACT

The transition to a future of 'Smart Mobility' – a mobility system characterised by real time organisation via the internet incorporating technologies such as connected and autonomous vehicles – has the potential to transform many aspects of everyday life. Many countries have evolved a system of 'multi-level governance' (MLG) to manage the formulation and implementation of public policies at different spatial scales. Whilst MLG has several potential advantages, such as providing multiple sites for policy innovation and de-risking the implementation of new policies by piloting them in particular places, the existence of many different governing tiers with different priorities and mandates requires skilful management and coordination. The management of any substantive, disruptive transition such as that to Smart Mobility is challenging for the policy system per se; for countries with MLG systems, the task is made more complex still by the need to achieve sufficient policy alignment between different tiers and entities of governance to implement new policy instruments in practice. The specific instruments of transport pricing and road-space reallocation provide clear examples of these challenges and pointers to how implementation questions might be resolved in an MLG framework.

Keywords: Transport; governance; multi-level; policies; alignment; action

Introduction

The transition to a future of 'Smart Mobility' – in essence the widespread application of a basket of technologies including real time sensing, continuous very high

Shaping Smart Mobility Futures:
Governance and Policy Instruments in times of Sustainability Transitions, 57–73

doi:10.1108/978-1-83982-650-420201004

speed internet connectivity and increasingly automation to the transport system so that the provision of mobility services becomes much more responsive to dynamic demand – is a major socio-technical transition (Geels, 2004, 2012) with the potential to transform many aspects of everyday life. However, as with any transition of this magnitude, the new end state made possible by these innovations, and the pathway to this future, are highly uncertain. The key posited benefits of Smart Mobility, primarily the much more efficient use of infrastructure and vehicles, combined with the application of robotics and artificial intelligence to create a safer, less wasteful and therefore lower carbon system of mobility, are only attainable if our systems of governance are able to manage and regulate these new technologies, their providers and the behaviour of users in the system effectively.

Many countries have evolved a system of 'multi-level governance' (MLG) to manage the formulation and implementation of public policies at different spatial scales (see Peters & Pierre, 2001; Piattoni, 2010; Pierre & Stoker, 2000; Scharpf, 1997). MLG can be understood as "the 'arrangement' of policy-making activity performed within and across politico-administrative institutions located at different territorial levels" (Stephenson, 2013, p. 817), or in more simple terms, the *distribution* of responsibility for the delivery of particular public policies between local-, regional-, national- and supra-national governments. In theory, MLG is a means of developing policies that are sensitive to the particular needs and characteristics of different places, whilst retaining a cohesive overall system of regulation and accountability that is clear and effective (see Charbit, 2011). If the system works well, MLG can provide a more varied set of opportunities for policy innovation to emerge (see Konvitz, 2016), specific interventions to be trialled or piloted in some places to minimise the risks involved in rolling out that innovation everywhere all at once, and the exchange of knowledge about 'what works' in practice between different governments and agencies.

But there are also difficult challenges involved in managing the interactions between tiers in an MLG system. Perhaps most important of these is the need to ensure that there remains sufficient *policy alignment* between the different tiers of government involved so that changes in policy direction can actually be achieved: Milio (2010) points to the potential for an 'implementation gap' to emerge if there is not careful management of political- and power relationships between institutions as there is always the potential for conflict between tiers to emerge and create barriers to effective policy delivery. Thus although when things go well, politics dictates that there will be competition to claim the credit, equally, when (new) policies fail to deliver, MLG relationships can quite quickly degenerate into a 'blame game' (Bache, Bartle, Flinders, & Marsden, 2015).

Transport is one of the domains of policy that has often been managed in a distributed manner in MLG systems. Responsibility for different aspects of the mobility system often rests with different tiers: standards, safety, regulation, taxation and 'strategic' or 'national' infrastructure at the top level in pursuit of common standards; some infrastructure, service planning, operation and subsidy at urban- or 'regional' level where perhaps differentiated fiscal responsibilities lie; and street management, enforcement and maintenance at 'local' or municipal level, where the delivery of services can be most immediately responsive to citizens. Supra-national regulation such as the EU's policies on engine standards adds yet another dimension to those policy fields such as climate change and economic regulation, where the problems (and solutions to them) are evidently larger than the capacity or jurisdiction of any single state.

The increasing prevalence of Smart Mobility technologies, and the promise of more fundamental change to mobility systems brought about by profound technological shifts such as connected and autonomous vehicles (CAVs) and the development of integrated 'Mobility as a Service' (MaaS) platforms offering real time, seamless multi-modal journeys (Smith, Sochor, & Karlsson, 2018), poses many difficult questions for any system of governance charged with managing the transition to Smart Mobility (Docherty, Marsden, & Anable, 2018). This is because the introduction of these technologies will require changes to a very large set of existing policy instruments, ranging from the law on road safety and crash liability to the framework for the taxation of vehicles, fuels and access to infrastructure, competition policy, operator licensing and the allocation of roadspace that are distributed across governance tiers (see case studies below). Given that the success of the Smart Mobility transition will depend in large part on how effectively each of these of these policy instruments is bundled together into a reformulated governance and regulatory environment that provides a stable context new forms of mobility to operate, the Smart Mobility transition is therefore arguably one of the most vibrant contemporary domains for MLG research to take place. Yet there is little if any research to date into how the governance institutions and processes responsible for framing and managing this transition will be constrained and/or empowered according to the governance model in place, let alone how the particular characteristics, advantages and disadvantages of MLG systems might influence the Smart Mobility transition.

This chapter therefore explores what will be required of the MLG systems now commonplace across Europe if they are to be able to manage the implementation of Smart Mobility technologies effectively. Building on previous research into how MLG systems have addressed decarbonisation and other pressing policy imperatives (see Bache et al., 2015; Bulkeley, 2005), the chapter uses the case studies of transport taxation and pricing, and road space management, to describe what the challenges to achieving sufficient policy alignment in order to support Smart Mobility might be. In particular, it analyses what difficulties might emerge in ensuring the degree of policy alignment between tiers required so that the system as a whole is able to formulate, agree and implement the new policy instruments required to make Smart Mobility innovations a reality.

The chapter proceeds as follows. After setting out the basic characteristics of MLG arrangements and the analytical frameworks used to approach them, it explores some key aspects of MLG systems in practice such as the need for actors to come together in policy 'moments' that generate debate about priorities and the potential to achieving alignment. It then discusses the implications of these processes of policy alignment for how the new, distinctive challenges that will be posed by the transition to 'Smart Mobility' might be articulated in different kinds of governing entities in an MLG system. Using two case studies of transport pricing and road space allocation, the analysis explores the extent to which the degree of policy alignment achieved between tiers will shape the wider system's capacity to cope with the pace of change as new mobility technologies become increasingly influential across the economy and society.

What is MLG and Why Does it Matter?

MLG analyses largely grew out of two parallel research concerns commonplace in the 1990s, namely exploration of the phenomenon of 'hollowing out' of the powers of the central state as the market became more involved in the provision of public services (see e.g. Rhodes, 1994; Skelcher, 2000), and the increase in activism and authority of supra-national entities, most importantly the European Union (Marks, Hooghe, & Blank, 1996; Peters & Pierre, 2001). More recent research on cities and city-regions has often stressed their (potential) ability to be more agile, innovative and entrepreneurial in many circumstances than national governments (see e.g. Konvitz, 2016; Pierre, 2019; Shearmur, 2012).

Fundamental therefore to the notion of MLG is that the power to act and implement policy no longer rests solely with the central state, and is instead to some degree *distributed* across a range of governing institutions that exist at different spatial scales, derive their legitimacy from different communities and interests, and are held accountable to different publics. The literature on the effectiveness of MLG systems in formulating and implementing public policy has therefore tended to focus on two critical concerns: the extent to which MLG systems are able to harness their potential flexibility to stimulate policy innovation and create the governing capacity to implement new policy instruments at pace in order to respond to socio-economic and technological change (Feiock, 2013) on the one hand, and how the potential for the complexities of the many interfaces of MLG systems to magnify inter-government competition and frustrate rather than support policy innovation (Ehnert et al., 2018) on the other hand.

In their article calling for a 'rethinking' of the now 20 year old MLG concept to acknowledge contemporary conditions, Alcantara, Broschek, and Nelles (2016, p. 39) argue that given the boundaries between the different entities that make up modern governance networks can be quite fluid, MLG is more than a means of understanding how complex government *systems* are structured; it is also an important analytical framework with which to analyse how the (changing) characteristics of the governance system determine the particular *instances* when key decisions – such as to implement a new policy instrument – come about at any particular point in time:

> At its simplest, MLG can be defined as an instance of policy-making in which government(s) engage with a variety of non-governmental actors, organized at different territorial scales, in a process of decision-making that aims to collaboratively produce some sort of public good.

As well as exploring how the *vertical* interfaces and relationships between different governments are mediated, Alcantara, Broschek, and Nelles go on to note how MLG models are able to incorporate analyses of *horizontal* linkages between different governance organisations active at a particular spatial scale, such as the city. MLG systems are therefore characterised by both spatial differentiation between formal tiers (e.g. municipality, region, and nation state) and also functional distinctiveness between different actors operating in any one of those spatial tiers. They point to Hooghe and Marks' (2003) categorisations of governing institutions according to whether they are multi-purpose or task specific in function. Multi-purpose institutions such as local- and regional governments that have a range of policy competencies can be regarded as 'Type 1' institutions, and exist to transfer the state's legal powers into the implementation of a 'national' policy responsibility such as education or healthcare at the subnational level. Then there is a much more diverse range of 'Type 2' or special purpose institutions – including non-departmental public bodies, other arms-length/special purpose government agencies, hybrid public–private partnerships and so on – that are the subject of much of the immense literature on the complexities of modern governance that has grown up in recent decades. This kind of institution, of which special purpose transport authorities are a prime example, is more likely to be focussed on specific policy domains or areas of technical expertise, and as such is often where the individuals with the greatest depth of technical knowledge and expertise on a particular policy domain or instrument are to be found in the wider governance network.

Thus in understanding the outputs of any MLG system of governance, it is crucial to explore both the 'vertical' distribution of power, agency and effectiveness between spatial scales, but also the 'horizontal' differentiation in knowledge, expertise, strategic capacity and power between different specialised bodies. As Bache and Flinders (2004) explore, the increasing prevalence of MLG systems has generated new lines of inquiry into how different institutions and actors develop their strategies for not only competition but also co-ordination and networking to safeguard their overall autonomy, and as a result, how the nature of democratic accountability of policy choices has evolved. For Smart Mobility, this means that analysis of policies aimed at facilitating and managing the introduction of new technologies in the transport domain need to take account of those in other, cognate areas as diverse as environmental policy, labour markets, planning and housing and indeed fiscal policy in general.

In their review of the implications of the shift to more complex and diffuse systems of governance, and within this general trajectory, the rise of MLG structures for transport and climate change, Marsden and Rye (2010) highlight the importance of a central implication of MLG, namely that the principal levers for the implementation of (new) policy instruments might be present at any of the spatial scales represented in the MLG structure, and therefore that understanding the potential for policy action in such a system requires both an appreciation of how the tiers interact, but also the unique political cultures of the territories and places represented by each MLG entity. Thus, as Paavola (2007, p. 93) noted in his consideration of how MLG has impacted on environmental governance, understanding the efficacy of these arrangements – and therefore how they generate the arrangement of actors in particular policy 'moments' that Alcantara, Broschek, and Nelles identified as crucial to actual policy choices – requires:

> analysis (that) can gain resolution by looking at the functional and structural tiers, organization of governance functions, and formulation of key institutional rules as key aspects of the design of governance institutions.

Given the above, it is perhaps unsurprising that a very significant proportion of the literature on contemporary governance structures, processes and outcomes has focussed on cities and city regions as the principal non-core state entities with the capacity to generate real policy change, given their scale, fiscal potential and long-standing political legitimacy (see e.g. Betsill & Bulkeley, 2007; Konvitz, 2016). Given their human and financial resources, city governments can play both Type 1 and Type 2 roles in an MLG system, and have the strategic capacity to lead policy innovation across several domains given that the task of urban governance is often centred on bringing stability and coherence to highly diverse places (Blanco, 2013; McCann & Ward, 2012). This is particularly true in the field of transport and mobility, given that many of the principal negative externalities of the transport system, such as congestion, local air pollution and (poor) service quality, are most often and immediately experienced at the urban level, and that efforts to 'solve' mobility problems by better integration between policies in transport, planning, economic development and so on has been most active at the urban level for many years (Marsden, Frick, May, & Deakin, 2011; Rode, 2019). Smart Mobility will be no different, because the enabling policy instruments it requires will also require coordination between the actions of a range of different governments both at different spatial scales but also between general- and special purpose authorities, for example, local municipalities with broad responsibilities and their own taxation powers, and transit agencies or development corporations.

(Dis)advantages of MLG

Such is the theoretical attractiveness of MLG given its potential to offer a genuinely pluralistic landscape for public administration, it has almost become accepted as a 'necessary' means of organising effective and accountable governance. As well as being a 'system' of governance according to its general definition, or a means of understanding how and why the particular policy 'moments' that create the scope to implement new policy instruments as set out by Alcantara, Broschek, and Nelles, MLG might also be argued to be a 'mindset'[1] of how the task of governing should be done to best effect. Yet according to two of the leading MLG scholars, 'beyond agreement that governance has become (and should be) multi-level,

---

[1]This very useful notion was suggested by Annica Kronsell at an authors' round table during the writing of this book.

there is no consensus about how it should be organized' (Hooghe & Marks, 2003, p. 233). This is perhaps because, despite its many attractive characteristics – MLG has also been associated with notions of localism (Stoker, 2004), 'stadspolitik' (i.e. a distinctive urban framework for governance reflecting the realities of city living; Dannestam, 2009), and in combination of all of the above, reinforced democratic legitimacy (see Papadopoulos, 2007) – every MLG structure is different, and the number of interfaces between different institutions in even a relatively simple governing network makes analysis challenging. But added to this technocratic complexity is the extra dimension of politics: no matter how complex the *administrative* relationships between different tiers and types of organisation, the *political* relationships across the MLG system, either in terms of hostility between different parties and interests in control of different elements or in the desire of those same parties to 'join up' efforts when they control, for example, entities in different tiers, can be more complex still. Thus, as Meadowcroft (2002, p. 171) counsels, it is 'important to appreciate just how much of political life cuts across the 'vertical' divisions of the formal hierarchy' in policy making.

Delegation, Gaming and the Tactics of Policy Making in MLG

Perhaps the best word to describe the challenge of achieving sufficient policy alignment in an MLG framework is therefore that it is 'messy' given the potential range of interactions between the different actors and institutions that together comprise a multi-level system in practice. But policy moments leading to change emerge from MLG systems nonetheless, and so in trying to explain how the complex processes and power dynamics of MLG work to produce real outcomes despite this 'sheer messiness', Chhotray and Stoker (2009) identified the key process of *delegation* between different governing organisations and territories as a lens to guide analysis, to which can be added the idea of different actors *gaming* the power- and political relationships between tiers to achieve desired outcomes.

The idea of delegation (at least implicitly) acknowledges that despite the complexity of MLG systems, the central state remains the critical local of power given its resources, legal competencies and role as the focus of political accountability and legitimacy. The success of policy initiatives in MLG – or what it is that makes policy 'moments' happen – is therefore in large part determined by how key actors choose to manage the power hierarchy between governing tiers. The most powerful tier or institution, usually central government – the 'boss' – has a strategic choice to make about whether it should either retain or delegate control over a particular policy responsibility or problem to one or more of its subordinate tiers. These other levels then have to decide whether to 'work' to achieve the goals set for them by the centre, or to 'shirk' responsibility, that is, to invite the centre to act itself if it thinks a particular problem important enough to warrant policy attention, or in some cases, simply to oppose and/or try to block central action if political disagreements between tiers are sufficiently stark.

Given the general enthusiasm for the idea, much MLG research focusses on the choices of central government over what to delegate, when and why, and how relationships between tiers are managed by those in the network. There are already several examples of delegation when it comes to policy instruments supporting Smart Mobility. For example, many city regions have been encouraged to develop their own frameworks for promoting Mobility as a Service (MaaS) platforms, including determining regulatory approaches to new market entrants such as Uber and Lyft in ridesharing, and dockless bikes/scooters such as Mobike and Lime.

The key to making MLG systems successful in their implementation of policy outcomes is therefore often conceptualised in terms of the *gaming* of the system. The idea of gaming can be understood as the efforts of principal actors in the governance network to create moments when sufficient policy alignment is achieved in order to take critical decisions. Doing this requires both active management of the often-conflicting priorities and relationships in a MLG system through dialogue, bargaining, and sometimes (political) compromise – and the continuous review of the participants and power distribution in an MLG system, so that the chances of achieving clear policy direction are maximised, known as *network management* (Klijn, Steijn, & Edelenbos, 2010).

But although work on delegation tends to be relatively optimistic about the capacity of MLG hierarchies to find a way through the management of difficult policy problems, it also comes up against some rather obvious criticisms about how well it really works in practice given the cut and thrust of politics in particular. In their work examining how different scales of governance on policies for decarbonisation, Bache et al. (2015) explained how MLG systems can quite easily break down into a 'blame game': if agreed policy targets, such as for emissions reduction are not met, then it is in the political interests of each governing entity to have others take and be seen to take responsibility for the failure. Thus, given that difficult policy challenges such as decarbonisation, or the effective implementation of Smart Mobility for that matter, require action by a range of organisations at different scales because of their very nature, even if an agreed set of delegated actions are agreed between actors can be established, actually implementing these agreed actions is always at risk from political 'events' that might have nothing to do with that particular policy domain. This danger, which in large part emerges from the fact that different tiers and types of organisation in an MLG system usually have different formal legal powers, constitutions, and electoral mandates and thus political imperatives, has led to deeper critiques about the sometimes ad hoc allocation of tasks in governing networks and the implications this might have for democratic accountability (see e.g. Sørensen & Torfing, 2009).

How will New Policy Instruments 'Test' MLG Arrangements?

The key research question for this chapter, which arises both from the Alcantara, Broschek, and Nelles and Bache, Bartle, Flinders, and Marsden analyses set out above, is therefore:

> how will MLG systems need to function so that sufficient policy alignment is achieved that those policy instruments required for the successful regulation and management of Smart Mobility are put in place?

We know from the broader empirical research base on MLG that successful policy alignment and therefore implementation of particular instruments requires negotiation between tiers and the development of a 'moment' in which it is in the interests of the majority of actors to pursue a particular path. But this is not the only 'test' that MLG systems (and the mindsets of those key actors within them) will face from the transition to Smart Mobility: in particular, the pace of technological development and (especially) the disruptive approach to regulation and compliance adopted by new entrants to the marketplace are already representing difficult challenges. What, therefore, are the best organising principles to achieve the necessary change in the overall set of public policy instruments that will be required to govern the Smart Mobility future well? And, how will MLG systems be able to do this at sufficient pace so that they keep up with technological progress and regulate powerful producer interests?

This is not to say that the existence of MLG is necessarily a constraint on the agility of the governance system: the explicit and effective delegation of responsibility to empower nominally 'lower' tiers of government such as cities to pursue more ambitious, rapid and comprehensive strategies for policy change is certainly possible. The crucial factor here is the particular of politics of transferring risk that accompanies delegation: if the actions of each tier are driven by the desire to externalise difficult and unpopular policy choices and decisions to other tiers rather than achieving policy alignment, then the kind of 'rhetoric: reality gap' that Marsden, Ferreira, Bache, Flinders, and Bartle (2014) identified between the bold and impressive policy statements about the need for governance to be genuinely joined up to achieve decarbonisation, and the reality of rather piecemeal policy implementation and real achievement on the ground will appear.

There are some clear lessons from these findings on MLG and decarbonisation (and indeed from similar studies in other policy domains such as innovation and regional economic competitiveness) for the good governance of the Smart Mobility transition. The first is that in any far reaching transition such as this, it is often the case that the *starting point* in terms of the distribution of powers and policy objectives prioritised by each tier is insufficiently aligned for the changes required to reach agreement and act to be easy or straightforward to achieve. One particularly important example for Smart Mobility that illustrates this well can be revealed by posing the apparently rather simple question of 'what is Smart Mobility actually for?' to different governments. At first glance, the answer for all institutions might be expected to be 'to make the transport system better'. But this is often not in fact the case: whereas some institutions do (of course) have this as their principal policy goal, others are actually more interested in creating a welcoming policy environment for Smart Mobility to encourage economic development and inward investment in relevant high technology sectors and/or wider urban competitiveness or marketing reasons: hence the tax breaks and designation of special 'future mobility zones' in the UK (DfT, 2019), the desire of some North American cities such as Pittsburgh and Toronto to pilot Uber's CAVs, and the (to this author at least) rather artificial construction of an 'Autonomous Vehicles Readiness Index' by one of the major global consultancy companies keen to extract value from country- and city authorities' desire to be seen as ahead of the pack in hosting high value blue chip companies (KPMG, 2019).

What Policies Need to be Aligned to Enable Smart Mobility?

In their review of what will be required for the good governance of Smart Mobility transition, Docherty et al. (2018) identify four key policy challenges that will need to be successfully negotiated and overcome to manage the transition positively in order to achieve increased public value from it. Each of these challenges presents its own test to MLG, and the extent to which its governing institutions and entities are able to manage the pressures of delegation and gaming so that they can create the policy moments that will lead to sufficient alignment to implement new policy instruments in practice.

Balancing the short versus the long game: the Smart Mobility transition implies that new sets of actors (such as the global technology companies and the data aggregators at the heart of Mobility as a Service models) will come to prominence as the principal agents in the wider mobility system. These actors will have their own (commercial) imperatives and, as ride sharing companies such as Uber have already done, will seek to challenge the existing 'rules of the game' on how mobility is regulated. One of the key tests for MLG here will be in how it is able to mediate the potential conflicts between the need for a 'comprehensive' reappraisal of regulatory frameworks at the upper national- and international tiers required for the provision of mobility in a new marketplace dominated by transnational corporations (Pangbourne, Stead, Mladenović, & Milakis, 2018), and the potential substantial disruption in some places that will come about as particular local markets adjust to specific changes brought about by Smart Mobility, such as the decline of the conventional bus beyond viability in response to the ubiquity of ridesharing (see Graehler, Mucci, & Erhardt, 2019). To address new forms of market failure and ensure public value, some form of 'comprehensive' (re)appraisal of overarching goals will be critical so that the contribution of new mobility options to the strategies of different MLG entities is evaluated and agreed. This will require real coordination between local and higher tiers so that overarching regulatory frameworks are designed in such a way that the actual level of service provision in different places meets the expectations of those places.

Who pays?: The envisaged shifts away from fossil fuels and towards a greater role for shared mobility will substantially change the core tax base for all governments with budgets that rely (at least in part) on revenues from mobility. Given the salience of these changes, this will likely require a comprehensive and wideranging debate about why and how mobility is taxed, and how future revenue raising can be done fairly and equitably. How then will MLG manage these potentially significant changes in the potential sources of revenue of tax revenues that will be available in the Smart Mobility future? Taxes on fuels are currently most often vested in central governments (with some exceptions, mainly in North America), so will the same be true if revenue instruments move towards per kilometre charges to reflect a changing structure of vehicle ownership/sharing and electrification in the Smart Mobility era? How will local demands to vary these charges according to particular local conditions such as the differential impact of congestion be accommodated? What will the links between revenue generation and the maintenance and enhancement of the infrastructure that supports the services that generate this revenue be?

Information asymmetries: public authorities of all kinds and at all spatial scales will need to be able to intervene effectively, so that private sector data hoarding does not lead to anticompetitive practices and/or generate the kinds of new market failures described above. Who will 'own' the data generated by the Smart Mobility system, and who has the right therefore to tax and/or use the data to support the management of mobility or other public services? Given one of the main purposes of the state is to act as an arbiter between actors with conflicting interests on standards and co-ordination, how will MLG set standards for data sharing and openness, and who will own the value generated by that data?

Equity and inclusion: We know that there is potential for Smart Mobility to both solve but also widen some equality and inclusion gaps according to how the transition plays out in practice (Jeekel, 2017). Ensuring social cohesion and inclusion is already a problematic area of activity for many MLG systems given the 'wicked' nature of the problem (Head, 2008), the need to align a wide variety of different policy instruments in different sectors in order to tackle it, and the tendency for lower tiers of government to take responsibility for the delivery of social care services. Following on from the challenges above, how will MLG redefine what a minimum standard of access to mobility might be in future, balancing profitable provision with social obligation? If a particular local jurisdiction wants to have higher standards of Smart Mobility provision for social inclusion or other reasons, how does it ensure this given the complexities of data control, revenue apportionment and other delegation issues outlined above?

Case Study Policy Instruments

In order to illustrate the kinds of policy alignment that will be required to implement the policy instruments necessary for the effective roll out of Smart Mobility, the final section of the chapter examines two case studies on key policy domains, where substantial policy redesign is likely to be required. In each case, the substantive issues of the delegation of policy responsibility, gaming the MLG system to create the potential for policy moments to emerge, the different attitudes of Type 1 and Type 2 institutions and attitudes to policy risk combine to frame the context for the delivery of new policy instruments for Smart Mobility in practice.

Transport Taxation and Pricing

Designing the optimal structure of taxation for mobility has long been a difficult task for governments of all types (see e.g. Mayeres, 2003). This is because there are many competing demands structuring the very definition of what 'optimal' in fact is for these purposes: on one level, there is the desire to price access to the transport network at a level that is optimised in terms of minimising purely mobility-related externalities such as congestion and pollution; then there is the desire to stimulate particular behaviours (e.g. in the labour market) that are shaped by choices in the wider taxation system; and then there is the simple need to generate revenue per se to support the complete basket of public services.

The first implication of Smart Mobility for taxation is simply that the scale of the behaviour change envisaged in the transition – from ownership to sharing of vehicles, and from discrete trip making to bespoke mobility packages determined in real time by MaaS, etc. – will require most systems of governance to reconstruct their systems of taxation for mobility to an extent that they have not done for decades. Given the scale of the tax revenue generated from hydrocarbons, and the fact that this revenue is usually tilted towards the higher tiers of government, it is not difficult to see how this challenge might be problematic, especially since as Reardon (2018) points out, the more centralised a policy function is, the more conservative the approach to policy change tends to be. Even opening up discussions about the future of mobility taxes at all represents a risk for those governments dependent on particular revenue streams.

How, then, will MLG systems address this change, and what are the implications for the actual policy instruments associated with transport pricing? As *The Oslo Study* on autonomous vehicles (COWI & PTV, 2019) notes, it is possible to design MaaS systems that have significant impacts on the overall size of the vehicle fleet and the distance operated by those vehicles. Thus when cities take the lead in designing and implementing MaaS systems, perhaps focussed on local policy objectives such as better traffic management and/or traffic reduction in city centres, or enhanced mobility for particular places or social groups in pursuit of particular inclusion and equality objectives, their decisions have a direct impact on the revenue streams that flow not only to themselves but also to other tiers of government. For example, given the size of the city within the overall Norwegian population, if anything approaching *The Oslo Study's* theoretical maximum 93% reduction in the number of vehicles were achieved in practice, then the central government would see its revenue streams from cars reduced dramatically. Equally, if central government were to impose a form of MaaS corresponding to one of the study's other scenarios in pursuit of wider economic or technology promotion objectives, then the city could be left with *double* the level of traffic and potentially little or no new revenue to try and mitigate this or even to maintain existing infrastructure given the scale of additional demand.

The disruptive potential of the Smart Mobility transition in terms of the *Who Pays?* questions of relative fiscal capacities and distribution of revenue vertically between governance tiers is therefore readily apparent, and this is even without addressing the implications of the electrification of the vehicle fleet, which is estimated to reduce the quantum of fossil fuels subject to taxation by up to a third in 15 years (Lindberg & Fridstrøm, 2015). Managing this transition will therefore require detailed, lengthy and complex negotiations between governance entities if a revised taxation structure for Smart Mobility is to be achieved that satisfies the key notions of fairness and effective allocation of risk. This is because there are several competing logics at play beyond the (relatively simple in theory, if not politically) question of the division of revenue between vertical tiers: the desire of (Type 2) transit agencies (often at the city- or regional scale) to ensure a system that maximises the performance of the transport system might not be entirely congruent with the desires of Type 1 entities that might have more complex demands to integrate mobility tax and pricing instruments with other fiscal tools and charges elsewhere in the economy, such as land and property taxes. Creating the policy 'moment' when a sufficient number of interests come together to generate negotiation, let alone policy alignment, might be extremely challenging. Finally, in larger countries with several cities seeking to manage the Smart Mobility transition, there exists the potential for a 'race to the bottom' in terms of which region can create the lowest tax regime in order to compete for investment. In such circumstances, higher tiers of government might wish to withdraw the delegation of some revenue powers so that the revenue accruing to the public sector as a whole at national level is safeguarded.

Roadspace Allocation

As road systems are hierarchical in nature, it is commonplace for different jurisdictions to manage different kinds of route: regional or federal governments manage the 'strategic' network of expressways and other major roads, with regional or urban governments managing intermediate routes, and municipalities responsible for local roads and streets. This means that the effects of the traffic management choices made by each kind of government spill over from one network to another. For example, the opening of a new strategic route will (dramatically) increase traffic levels on the local route network at either end; equally, network management decisions such as congestion pricing (see above) can divert vehicles away from urban centres onto strategic routes and vice versa.

There are, therefore, a number of first order effects from decisions about how to manage roadspace, such as congestion moving from one network to that in another jurisdiction because a municipality has introduced dedicated public transport lanes on key routes. These issues can be both vertical and horizontal in terms of how they affect the level of service and policy objectives for mobility across the entities making up an MLG system. As *The Oslo Study* once again demonstrates, the nature of these decisions in the era of Smart Mobility will have many consequences beyond that of determining the absolute level of traffic, especially in cities where transport systems can have crucial structuring issues on the land and property markets, employment and the potential for wider economic benefits. For example, detailed decisions about how to implement MaaS and ride sharing – particularly those about which parts of the city ridesharing vehicles can operate in, and the regulatory rules about how many vehicles are permitted in the ridesharing pool, and the extent to which this pool of shared vehicles has preferential access to roadspace such as dedicated lanes – will have major implications for the performance of the wider transport network, and which kind of journeys are privileged in terms of time, cost and convenience.

In this context, Smart Mobility might lead to renewed conflicts between the interests of different governments that will need to be opened up, negotiated and agreed across the MLG system. City governments that take the decision to 'go for' MaaS at scale as a means to improve the performance of their transport system (and economy more widely) will therefore need to use all the new policy instruments at their disposal to make ridesharing as attractive as possible. This will mean (quite radical) reallocation of roadspace to shared vehicles and away from single occupancy cars, rethinking parking strategies, and even rewriting the instruments that determine which vehicles have access to curbspace, given that the attractiveness of ridesharing in large part depends on its ability to deliver genuinely door-to-door journeys.

Marsden, Docherty, and Dowling (2020) explore how the curbside of the future will be the site of tensions between competing interests not just between different user groups – drivers, passengers, cyclists, pedestrians – but also the governance entities that traditionally represent them. Even more than for space on the road, the use of space on the curb has long been tightly regulated by municipalities through stopping and waiting restrictions and user charging. How access to the curb is managed in future will be a significant determinant of the performance of many Smart Mobility systems: for example, if CAVs are restricted to relatively few designated drop off and pick up zones then their attractiveness compared to traditional public transport services might be substantially reduced. Equally, given the now quite substantial and long standing trend to the de-motorisation and pedestrianisation of city neighbourhoods, there is potential for a clear disjuncture between the objective of national governments to meet climate change obligations by promoting MaaS and shared services as much as possible, and the real world local management of the impacts of these services on local communities and firms, and on other users and Smart Modes such as bike share services, and pedestrians.

Given what we know about the politics of roadspace reallocation – which often enters popular debate as an 'attack on the motorist' – the at first seemingly rather mundane questions about how to plan for new Smart modes so that they offer efficient mobility options in practice are in fact potentially quite problematic on deeper inspection. The (re)allocation of roadspace to different Smart Mobility modes, and therefore the utility of these modes to their users and citizens more generally, is an intensely political issue not just because of its impact on everyday life, but because the local tiers of an MLG system often derive much of their democratic legitimacy from their ability to manage services and the public realm at the micro level. Thus many of the meta- or strategic policy instruments that might be created by higher tiers of government to support Smart Mobility – such as the rules about the interaction between CAVs and pedestrians and whether there should be 'designated driverless spaces' to encourage the uptake of the technology – might have profound impacts on the lived experience of places for many people. To date, given the normative zeal that many governments have for the introduction of Smart Mobility, there has been relatively little debate about these effects. But once the scale of the transition that the built environment, public realm and spatial organisation of the economy that innovations such as CAVs will bring becomes sufficiently understood, the task of aligning regulatory frameworks and everyday management of the system might be very challenging indeed.

Conclusions

The potential advantages of MLG as an organising concept for policy action hinge around its potential to match strategic decision making with detailed empirical knowledge about the implications of implementing specific policy instruments at different scales and in real places. As the various modelling efforts of the impacts of autonomous vehicles following the original International Transport Forum study on Lisbon demonstrate (ITFCPB, 2017), the actual *outcomes* in terms of what kind of network, level of service and equity of access Smart Mobility technologies will bring is highly dependent on the precise choices that are made in terms of its implementation, and the impacts of local factor conditions such as urban morphology and the characteristics of the local public transport system. If Smart Mobility is to be successful, then the potential of MLG systems to manage the transition by tailoring particular the policy instruments required – such as the pricing and space allocation cases above – will have to be realised otherwise negative externalities might solidify into real barriers to implementation.

Further, the impacts of transport choices have many different intended and unintended consequences, some positive, some negative, and even if there is substantial consensus that a policy 'moment' in which the need for action has arrived, these complexities mean that actually achieving that alignment so that new policy instruments can be agreed and implemented is by no means inevitable. Doing so is a stiff test of what is often called 'statecraft' when considering implementation pathways at national level, with the added complexity of multiple levels of governing 'craft' to manage. As for so many analyses of Smart Mobility therefore, focus on MLG shows that it is not the technological challenges that are likely to frustrate its adoption, but the socio-economic and political ones where greatest complexity lies.

References

Alcantara, C., Broschek, J., & Nelles, J. (2016). Rethinking multilevel governance as an instance of multilevel politics: A conceptual strategy. *Territory, Politics, Governance*, *4*(1), 33–51.

Bache, I., Bartle, I., Flinders, M., & Marsden, G. (2015). Blame games and climate change: Accountability, multi-level governance and carbon management. *The British Journal of Politics and International Relations*, *17*(1), 64–88.

Bache, I., & Flinders, M. (2004). Multi-level governance and the study of the British state. *Public Policy and Administration*, *19*(1), 31–51.

Betsill, H., & Bulkeley, H. (2007). Rethinking sustainable cities: Multilevel governance and the 'urban' politics of climate change. *Environmental Politics*, *14*(1), 42–63.

Blanco, I. (2013). Analysing urban governance networks: bringing regime theory back in. *Environment and Planning C: Government and Policy*, *31*(2), 276–291.

Bulkeley, H. (2005). Reconfiguring environmental governance: Towards a politics of scales and networks. *Political Geography*, *24*(8), 875–902.

Charbit, C. (2011). *Governance of public policies in decentralised contexts: The multi-level approach*. Paris: OECD.

Chhotray, V., & Stoker, G. (2009). *Governance theory and practice: A cross-disciplinary approach*. Basingstoke: Palgrave Macmillan.

COWI & PTV. (2019). *The Oslo study – How autonomous cars may change transport in cities*. Oslo: Ruter. Retrieved from http://www.mynewsdesk.com/no/ruter/documents/the-oslo-study-how-autonomous-cars-may-change-transport-in-cities-87313. Accessed on December 17, 2019.

Dannestam, T. (2009). Stadspolitik i Malmö. Politikens meningsskapande och materialitet [City politics in Malmö: Political meaning creation and materiality], Lund Political Studies 155. Retrieved from https://portal.research.lu.se/portal/sv/publications/stadspolitik-i-malmoe(dd547b26-a9f5-4131-83ca-5e0e67087611).html.

DfT. (2019). *Future mobility zones fund: Competition process and shortlisted applicants*. London: DfT. Retrieved from https://www.gov.uk/government/publications/future-mobility-zones-fund-competition-process. Accessed on November 25, 2019.

Docherty, I., Marsden, G., & Anable, J. (2018). The governance of smart mobility. *Transportation Research Part A: Policy and Practice*, *115*, 114–125.

Ehnert, F., Kern, F., Borgström, S., Gorissen, L., Maschmeyer, S., & Egermann, M. (2018). Urban sustainability transitions in a context of multi-level governance: A comparison of four European states. *Environmental Innovation and Societal Transitions*, *26*, 101–116.

Feiock, R. C. (2013). The institutional collective action framework. *Policy Studies Journal*, *41*, 397–425.

Geels, F. W. (2004). From sectoral systems of innovation to socio-technical systems: Insights about dynamics and change from sociology and institutional theory. *Research Policy*, *33*(6–7), 897–920.

Geels, F. W. (2012). A socio-technical analysis of low-carbon transitions: Introducing the multi-level perspective into transport studies. *Journal of Transport Geography*, *24*, 471–482.

Graehler, M., Mucci, R. A., & Erhardt, G. D. (2019). Understanding the recent transit ridership decline in major US cities: Service cuts or emerging modes?. In *98th Annual Meeting of the Transportation Research Board (TRB)*, Washington DC, USA.

Head, B. W. (2008). Wicked problems in public policy. *Public Policy*, *3*(2), 101.

Hooghe, L., & Marks, G. (2003). Unraveling the central state, but how? Types of multi-level governance. *American Political Science Review*, *97*, 233–243.

International Transport Forum Corporate Partnership Board. (2017). *Transition to shared mobility: How large cities can deliver inclusive transport services*. Paris: OECD-ITF Corporate Partnership Board Policy Insights, June.

Jeekel, H. (2017). Social sustainability and smart mobility : Exploring the relationship. *Transportation Research Procedia*, *25*, 4296–4310.

Klijn, E., Steijn, B., & Edelenbos, J. (2010). The impact of network management on outcomes in governance networks. *Public Administration*, *88*(4), 1063–1082.

Konvitz, J. (2016). *Cities and crisis*. Manchester: Manchester University Press.

KPMG. (2019). *2019 autonomous vehicles readiness index*. Zurich, Switzerland: KPMG International *Cooperative*. Retrieved from https://assets.kpmg/content/dam/kpmg/xx/pdf/2019/02/2019-autonomous-vehicles-readiness-index.pdf. Accessed on November 25, 2019.

Lindberg, G., & Fridstrøm, L. (2015). *Policy strategies for vehicle electrification*. ITF/OECD Discussion Paper 2015:16, ITF/OECD, Paris, France.

Marks, G., Hooghe, L., & Blank, K. (1996). European integration from the 1980s: State-centric v. multi-level governance. *Journal of Common Market Studies*, *34*(3), 341–378.

Marsden, G., Docherty, I., & Dowling, R. (2020). Parking futures: Curbside management in the era of 'new mobility' services in British and Australian cities. *Land Use Policy*, *91*, 104012..

Marsden, G., Ferreira, A., Bache, I., Flinders, M., & Bartle, I. (2014). Muddling through with climate change targets: A multi-level governance perspective on the transport sector. *Climate Policy*, *14*(5), 617–636.

Marsden, G., Frick, K. T., May, A. D., & Deakin, E. (2011). How do cities approach policy innovation and policy learning? A study of 30 policies in Northern Europe and North America. *Transport Policy*, *18*(3), 501–512.

Marsden, G., & Rye, T. (2010). The governance of transport and climate change. *Journal of Transport Geography*, *18*(6), 669–678.

Mayeres, I. (2003). Taxes and transport externalities. *Public Finance and Management*, *3*(1) 94–116.

McCann, E., & Ward, K. (2012). Assembling urbanism: Following policies and 'studying through' the sites and situations of policy making. *Environment and Planning A*, *44*(1), 42–51.

Meadowcroft, J. (2002). Politics and scale: Some implications for environmental governance. *Landscape and Urban Planning*, *61*(2–4), 169–179.

Milio, S. (2010). *From policy to implementation in the European Union: The challenge of a multi-level governance system*. London: Tauris Academic Studies.

Paavola, J. (2007). Institutions and environmental governance: A reconceptualization. *Ecological Economics*, *63*(1), 93–103.

Pangbourne, K., Stead, D., Mladenović, M., & Milakis, D. (2018). The case of mobility as a service: A critical reflection on challenges for urban transport and mobility governance. In G. Marsden & L. Reardon (Eds.), *Governance of the smart mobility transition* (pp. 33–48). Bingley: Emerald.

Papadopoulos, Y. (2007). Problems of democratic accountability in network and multilevel governance. *European Law Journal*, *13*(4), 469–486.

Peters, B., & Pierre, J. (2001). Developments in intergovernmental relations: Towards multilevel governance. *Policy and Politics*, *29*(2), 131.

Piattoni, S. (2010). *The theory of multi-level governance: conceptual, empirical, and normative challenges*. Oxford: Oxford University Press.

Pierre, J. (2019). Multilevel governance as a strategy to build capacity in cities: Evidence from Sweden. *Journal of Urban Affairs*, *41*(1), 103–116.

Pierre, J., & Stoker, G. (2000). Towards multi-level governance. In P. Dunleavy, A. Gamble, I. Holliday, & G. Peele (Eds.), *Developments in British politics* (6th ed., pp. 29–46). London: Macmillan.

Reardon, L. (2018). Networks and problem recognition: Advancing the multiple streams approach. *Policy Sciences*, *51*(4), 457–476.

Rhodes, R. (1994). The hollowing out of the state: The changing nature of the public service in Britain. *The Political Quarterly*, *65*(2), 138–151.

Rode, P. (2019). Urban planning and transport policy integration: The role of governance hierarchies and networks in London and Berlin. *Journal of Urban Affairs*, *41*(1), 39–63.

Scharpf, F. (1997). Introduction: The problem-solving capacity of multi-level governance. *Journal of European Public Policy*, *4*(4), 520–538.

Shearmur, R. (2012). Are cities the font of innovation? A critical review of the literature on cities and innovation. *Cities*, *29*(Supp 2), S9–S18.

Skelcher, C. (2000). Changing images of the state: Overloaded, hollowed-out, congested. *Public policy and Administration*, *15*(3), 3–19.

Smith, G., Sochor, J., & Karlsson, M.-A. (2018). Mobility as a service: Development scenarios and implications for public transport. *Research in Transportation Economics*, *69*, 592–599.

Sørensen, E., & Torfing, J. (2009). Making governance networks effective and democratic through metagovernance. *Public Administration*, *87*(2), 234–258.

Stephenson, P. (2013). Twenty years of multi-level governance: 'Where does it come from? What is it? Where is it going?'. *Journal of European Public Policy*, *20*(6), 817–837.

Stoker, G. (2004). New localism, progressive politics and democracy. *The Political Quarterly*, *75*, 117–129.

Chapter 5

Planning Urban Futures for Autonomous and Shared Vehicles: The Role of Planning Support Tools as a Policy Instrument

Sam McLeod, Carey Curtis and John Stone

ABSTRACT

Modelling has been a mainstay of conventional planning support tools (PSTs) since the 1960s and is instrumental in transport and land use planning decision-making. Numerous studies have been conducted to model the potential impacts of emerging vehicle automation and sharing technologies. A systematic review of recent modelling studies of autonomous and shared vehicles in the research literature examines the extent of their contribution to 'smart' mobility knowledge. The findings suggest a limited knowledge base from which to support future planning. PSTs that can offer more pluralistic, discursive, and transparent methods in order to understand and proactively shape a transition to a planned urban future are also needed.

Keywords: Autonomous vehicles; planning support tools; transport modelling; strategic planning; urban planning; mobility as a service

Introduction

For many years, the idealised task of transport planners and engineers has been to identify ways that urban transport and land-use systems can be shaped by regulation and investment to meet environmental, social and economic objectives. A large part of this task involves describing and analysing the hugely complex political and technical dynamics of the urban transport system and applying this analysis to help politicians and citizens to make informed decisions about future actions.

Shaping Smart Mobility Futures:
Governance and Policy Instruments in times of Sustainability Transitions, 75–103

doi:10.1108/978-1-83982-650-420201005

Simply stated, the planning process works to bring interested parties together to identify shared values and objectives and to define and implement strategies and actions that can meet those objectives. This is necessarily a discursive process. Many techniques – more recently called 'Planning Support Tools' (PSTs) – have been developed to help participants to visualise complexity and to imagine pathways to desired outcomes. In practice, these processes are shaped by imbalances in power and access to participation, and the legitimacy of outcomes is often contested.

Deeply embedded in formal transport and land-use planning processes is a reliance on mathematical models. Although often internally complex, these models are ultimately only simplified representations of urban system dynamics that, in their best use, allow for exploration of the impacts on outcomes such as traffic volumes, air pollution, or public transport access from changes to infrastructure or service patterns (Gudmundsson, 2011). As such, they can make an important contribution to a discursive process. However, the complexity of many of these models leads to some serious problems (Lee, 1973; Saujot, de Lapparent, Arnaud, & Prados, 2016). The technicalities of the modelling can become an end in themselves, with many modellers losing sight of the important truism[1] that: 'all models are wrong, but some are useful'. This is compounded by the desire of many decision-makers for a simple numerical 'answer' that can be used to justify a pre-determined outcome (Næss, 2006). The opportunities for distorted interpretations of modelling outputs are exacerbated by the distance that typically lies between practitioners with detailed understanding of the assumptions and limitations of a particular model and decision-makers who ultimately use the outputs (Fig. 1).

For these reasons, a great deal of care is required in the interpretation of published conclusions drawn from a transport model. It is necessary to ask: Who commissioned the model and directed its specifications? What questions is it asking and what assumptions have been made? Who is receiving the results and how are they using them in current discourse?

This need for care is doubly important when models are attempting to shed light on transport patterns in a future in which new technologies for mobility are being deployed. The uncertainties surrounding many questions, especially with respect to public preferences towards vehicle ownership and in-vehicle sharing, underline the dangers of over-reliance on 'headline' findings. An array of knowledge is required by urban planners, including: which journey purposes will be served by these modes; what the take up of these modes will be by demography; and what impacts on the road and street infrastructure, parking, congestion and on urban structure may be (Curtis, Stone, Legacy, & Ashmore, 2019). Transition to an Autonomous Vehicle (AV) 'end-state' will take decades, and the progression through the evolution of the vehicle fleet and the inevitable mixing of different stages of the technologies will be complicated, adding to uncertainty in the extent and speed of transition.

[1]Generally attributed to statistician George E. P. Box.

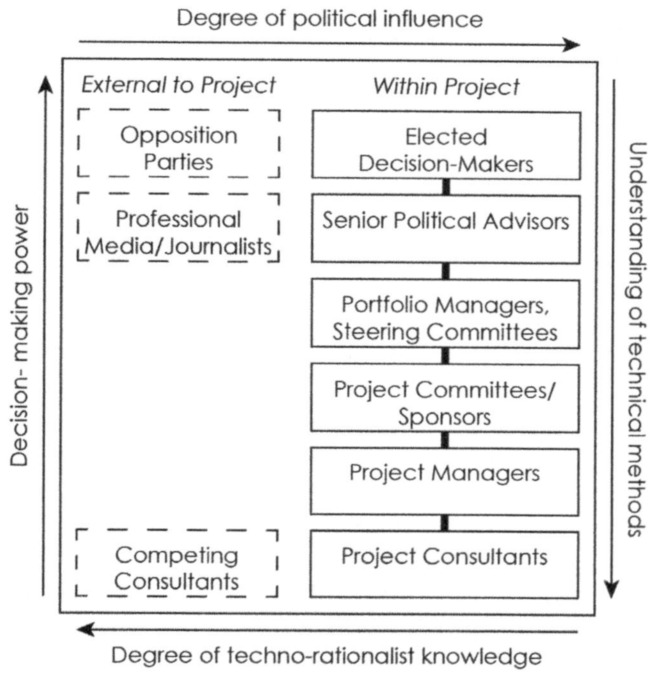

Fig. 1. Typical Organisation of a Transport Planning Project and the Influences upon it. *Source*: Adapted from Evans, Burke, and Dodson (2007).

In this chapter, we ask these questions of a large sample of the published model outputs available up to 2018 – these represent a rapid rise of research in forecasting and modelling the implications of AVs. After reviewing the content and limitations of these studies, we return to the question of what tools are most useful in planning for sustainable transport outcomes in the age of new technologies. This refers attention back to the need for PSTs that can offer more pluralistic, discursive, and transparent methods.

Approach

This analysis draws on a systematic literature review (Van Wee & Banister, 2016) of journals in the fields of transportation, urban studies, planning, and geography, supplemented with searches of Scopus and Google Scholar. A database with fields for mobility type (Table 1), modelling technology, geographical context, sample/input data characteristics, and thematic findings was completed for each publication. Forty nine publications were assessed as presenting peer reviewed modelling research having direct relevance to our inquiry.

Table 2 lists the identified corpus of literature, grouped by geography and then by research group, in order to illustrate multiple publications reporting similar research. Generally, models in each group have been iteratively developed,

Table 1. Mobility Type Definitions.

Application Type	Definition
PCV	Private Conventional Vehicle (human driver, Level 0 automation)
AV	Autonomous vehicle without specific definition of ownership (i.e. as described generally or for studies which did not relate to supply/demand dynamics)
EV	Electric vehicle; irrespective of form of automation or ownership
Private AV	AVs are privately owned and used only by owners/ household
Shared AV	Autonomous taxi-services, used by any member of the public
Dynamic Ride Shared AV	Autonomous taxi which may concurrently serve multiple users with the same or closely located trip origin and destination points
Levels 0–5	As per Society of Automotive Engineers definitions

particularly as dissertations and pilot projects have transferred into full peer-reviewed publication, and as models are expanded for larger geographic areas or built to represent different elements of transport systems. Where this occurred, input datasets tend to be very similar as research progresses, particularly as synthetic populations or trip demand sets are re-used.

'Smart' Mobility Knowledge and Limitations of Modelling Analyses

The scope of AV modelling is extremely broad; including a range of aspatial analyses (such as purely theoretical road network capacity assessments or single link/ intersection studies). Publications generally profiled the development and technical performance of a model, without describing use in a specific planning process or exploration of the results with policy-makers; while this does not imply the model is used in planning practice, it cannot be ruled out. Some authors did stake claim to the value of their tools in policy-making or in informing policy-makers (Hawkins & Nurul Habib, 2018; The Boston Consulting Group, 2017; Zhang, Guhathakurta, & Khalil, 2018), although consistent with Marsden and Reardon (2017), most papers lacked clear policy articulation.

Table 2 indicates a diversity of purpose and underlying funding of studies. Only a minority appear to have been commissioned directly by a planning agency with respect to a specific policy issue; rather, most studies appear to be preliminary in nature, typically enabled through scholarships or through research institute funding. Measuring the precise extent of influence these modelling studies is not possible – though citation counts provide one indicator of wider reach.

Table 2. Publications Reviewed by Geography and Mode.

Reference	Model Geography	Vehicle Type/s	Assumptions Applied	Commissioned By/Collaboration With	Citation Record[a]
United States					
Austin, Texas Group (Fagnant, Kockelman et al.)					
Fagnant and Kockelman (2014)	Hypothetical grid (10 mile square)	SAV	Fixed link speeds; waiting times, 3.5% of total trips served by SAVs, no multi-occupant sharing	Not stated	518
Fagnant, Kockelman, and Bansal (2015)	Austin, Texas (subset area only)	SAV	A randomly-selected 5% of total trip volume modelled, then 100,000 personal trips with destination within central area (12-mile by 24-mile 'geofence') assumed to use SAVs, each simulated vehicle represents approximately 20 cars, all other demand assumed to use other modes	Not stated; uses CAMPO metropolitan planning agency trip tables	182
Chen, Kockelman, and Hanna (2016)	Simple 100 by 100 mile four concentric zone grid city modelled on Austin, Texas	SAV (all EVs)	Range and vehicle charging time, some costs, constant travel speed	Scholarship funded by the National Science Foundation	152
Chen and Kockelman (2016)	Grid city modelled of Austin, Texas	SAV (all EVs)	Same fixed speed and static trip generation rates as previous research. Non-motorised trips not modelled. Income randomly assigned to agents. Assumes transit is 20–25% slower than fixed SAV speeds, fixed transit wait times. Many minor assumptions based on existing data for Austin or US	Same as above	76
Liu, Kockelman, Boesch, and Ciari (2017)	Austin, Texas (detailed city-wide road network)	SAV	Assumes transit trips do not have transfers; Value of Travel Time (VOTT) assumptions for modes, acknowledges destination choices would likely be different if AVs were available	Funded by the Texas Department of Transportation (TxDOT)	36

Table 2. (*Continued*)

Reference	Model Geography	Vehicle Type/s	Assumptions Applied	Commissioned By/Collaboration With	Citation Record[a]
Loeb, Kockelman, and Liu (2018)	Austin, Texas (detailed city-wide road network)	SAV (EVs)	5% of total trips modelled, assume 2% of these use SAEVs (0.1% overall mode share). Takes some SAEV assumptions from (Chen, Kockelman, and Hanna 2016)	Same as above	33
Fagnant and Kockelman (2018)	Austin, Texas (subset area)	SAV, DRS	Random subset of trips drawn, only modelled within 12-mile by 24-mile 'geofence' – trips outside this assumed to be by other modes	Not stated; uses CAMPO metropolitan planning agency trip tables	154
Zhao and Kockelman (2018)	Austin, Texas region	PAV, SAV	2020 AM peak forecast. Mode choice model simplified to only: Car, PAV, SAV, and Bus. VOTT and operating costs assumed. Trucks included in model but not in reported results. Hourly link speeds	Funded by the Texas Department of Transportation (TxDOT); uses CAMPO metropolitan planning agency trip tables	12
Cell Transmission Models (Levin et al. Group)					
Levin and Boyles (2015)	Austin (downtown network not further described)	PAV	Driving or transit mode choice (no active mode complete trips), static traffic assignment, fixed fuel cost. On-road transit only (no exclusive right of way), link capacity correlated to proportion of AVs	Supported by Data-Supported Transportation Operations and Planning (D-STOP) University Transportation Centre	66
Patel, Levin, and Boyles (2016)	Very small sub-network areas in Austin, Texas	AVs	Reduced reaction times of AVs, reservation-based intersections are possible within urban areas	Supported by TxDOT, D-STOP, and the National Science Foundation	15
Levin, Kockelman, Boyles, and Li (2017)	Austin, Texas (downtown area specified)	PAV, SAV, DRS	Assumes no private vehicles, fixed maximum waiting time, intra-zone trips excluded	Same as above	86

Chicago (Auld et al.) Group					
Auld, Sokolov, and Stephens (2017)	Chicago, Illinois, road and public transport networks	PAV	Assumes no empty running, assumes conventional intersection control could be removed, assumed consistent spatial diffusion of AV ownership	Not stated, work based on existing regional model	22
Auld, Verbas, Javanmardi, and Rousseau (2018)	Chicago, Illinois	PAV (level 4)	Assumed marginal cost of technology and willingness to pay (WTP) distribution, assumed lower VOTT, link capacity increases proportional to AV share	Funded by US Department of Energy Vehicle Technologies Office	1
Boston (BCG) Group					
The Boston Consulting Group (2017)	0.45 km^2 area of central Boston	PAV, SAV	Assumes no walking or cycling, other modal shares assumed, assumed AV driving characteristics, willingness to share	BCG and WEF joint initiative involving testing AEVs on ground. Initiative designed by 'city leaders'. City of Boston, DoT. MIT Media Lab. NuTonomy. Abertis, Audi, BMW, Chargepoint, City of Gothenburg, Denso, Ericsson, General Motors, Honda, Hyundai Motor Company, Innogy, Jaguar, Land Rover, Land Transport Authority Singapore, Liberty Mutual;, Lindholmen Science Park – Drive Sweden, Lyft, MoT Singapore, Optimus Ride, Qualcomm, Renault-Nissan Alliance, Robert Bosch, Sienens, SOMPO Holdings, State Farm, Toyota Research Institute, Transdev, UPS, Volkswagen Group, Volvo Group	10

Table 2. (*Continued*)

Reference	Model Geography	Vehicle Type/s	Assumptions Applied	Commissioned By/Collaboration With	Citation Record[a]
World Economic Forum (2018)	Entire City of Boston area (~145 km^2)	SAV, DRS	Capacity assumptions derived from single link microsimulation	BCG, WEF, City of Boston. Participation of Autonomous and Urban Mobility Working Group: Abertis, Airbus, Audi Boston Transportation Dept., Bridgestone, ChargePoint, City of Gothenburg, Denso Corp., Deutsche Post DHL, Drive Sweden, Enel, Engie Group, Ericsson, General Motors, Guangzhou Automobile Group, Hyundai, Innogy, Inrix, Jaguar Land Rover, Keolis, LTA Singapore, Lyft, nuTonomy, Renault-Nissan-Mitsubishi, Robert Bosch, Saudi Aramco, SNC-Lavalin, Sompo Holdings, Toyota Daihatsu Engineering, Toyota Research Institute, US Office of Governor of Washington, UPS, Volkswagen, Volvo, Zurich Insurance Group	No data
New Jersey (Kornhauser et al.) Group					
Zachariah, Gao, Kornhauser, and Mufti (2014)	Grid representation of New Jersey	SAV, DRS	Taxi stands used for start and end points, one stand at centroid of each raster area. No network (Manhattan distances used)	Sponsored by TRB committee AP060 Paratransit	29
Brownell and Kornhauser (2014)	Grid representation of New Jersey	SAV, DRS	Assumed travel speeds, capital cost per vehicle	Not stated	51

Atlanta (Zhang et al.) Group					
Zhang, Guhathakurta, Fang, and Zhang (2015)	Hypothetical city	SAV, DRS	2% of population used SAVs, constant link speeds	Not stated	122
Zhang and Guhathakurta (2017)	Atlanta, Georgia	SAV, DRS	5% SAV market penetration	Not stated	28
(Zhang et al., 2018)	Atlanta, Georgia	PAV	No induced demand. AVs shared only within single household. Replacement of private cars with PAV at household level. Number of AV owned based on peak household demand	Not stated; use of Atlanta Regional Commission data	32
Miscellaneous Other Studies					
Shen and Lopes (2015)	New York City	SAV	Randomly generated passenger waiting tolerances, AV speed limit	Collaborative use of *MobilityTestbed* and Agent-Polis platform	27
Childress, Nichols, Charlton, and Coe (2015)	Seattle, Washington (Puget Sound region)	PAV, SAV	Mode and trip choice model – network performance evaluation method not clear (likely EMME). Uses 2010 base year	Not stated	137
Hörl, Erath, and Axhausen (2016)	Sioux Falls	SAV	VOTT (work and non-work trips), costs for modes	Not stated	10
Masoud and Jayakrishnan (2017)	San Diego County, California	SAV	Households grouped into clusters which then share SAVs	Not stated	12
Mendes, Bennàssar, and Chow (2017)	Single line (proposed Brooklyn-Queens light rail), New York	DRS	Assumed DRS fleet operates as alternative to light rail along project alignment. Induced demand assumed for project case	Supported by the New York University Undergraduate Summer Research Program; Fulbright scholarship	8

Table 2. (*Continued*)

Reference	Model Geography	Vehicle Type/s	Assumptions Applied	Commissioned By/Collaboration With	Citation Record[a]
Merlin (2017)	Ann Arbor, Michigan	SAV, DRS	Maximum 10 minute inconvenience for detours to pick up additional riders, assumed capacity, fixed speed, fleet size set by max average waiting time	Not stated	20
Europe					
Dutch Case Studies (Correia et al.) Group					
Winter, Cats, Correia, and Arem (2016)	Single line with only two destinations, Wageningen, Netherlands	DRS	Vehicles are electric, maximum waiting time guaranteed	Part of the WEpods Project; Openbaar Vervoer-Oost provided current demand data.	17
Correia and van Arem (2016)	Simplified 61 link, 46 node network of Delft, the Netherlands	PAV	152 trips (29 households) modelled used to represent thousands based on a synthetic sampling expansion factor. Travel within zones assumed to be walking. No parking limitations. Assumes all modelled demand uses PAVs or public transport	Partly funded by the WEpods project (Province of Gelderland) and the D2D100% EV vehicle automation pilot projects in the Netherlands	77
Liang, Correia, and van Arem (2016)	Zuid Delft station, the Netherlands	SAV (EV)	Random assignment of trips to SAVs. Single-seat vehicles. Cost and performance assumptions. Static routes between fixed origin and destination zone centroids	D2D100%EV funded by the Dutch rail infrastructure manager ProRail; WePods project (Province of Gelderland); Chinese Scholarship Council	35

Munich SITO-MITO-MATSim (Moreno et al.) Group					
Llorca, Moreno, and Moeckel (2017)	Greater Munich metropolitan area	SAV, PCV	Mode selection according to distance, random assignment between conventional vehicles and SAVs. Random destination assignment. Public transport model logic not defined	Supported by the German Excellence Initiative and the European Union Seventh Framework Programme under grant agreement 291763; the TUM University Foundation Fellowship (TUFF) for international postdocs; funded by the Deutsche Forschungsgemeinschaft (DFG, German Research Foundation), Project MO 1420/2-1	3
Moreno, Michalski, Llorca, and Moeckel (2018)	Same as above	SAV	Similar to above. SAV mode share based on survey results.	Supported by the Technische Universität München-Institute for Advanced Study, funded by the German Excellence Initiative and the European Union Seventh Framework Programme under Grant Agreement no. 291763. Assistance from the Department of Transport Planning and Mobility Management in the city of Fürstenfeldbruck	5
OECD/ITF (Martinez et al.) Lisbon Group					
OECD/ITF (2015)	Lisbon municipality	SAV, DRS	Replace all car and bus trips with SAV/DRS trips. Hourly link speeds. Three AV vehicle capacities. Trip characteristic constraints	Initiated and funded by the International Transport Forum's Corporate Partnership Board (CPB). Participating Corporate Partners – Michelin and Nissan	33

Table 2. (*Continued*)

Reference	Model Geography	Vehicle Type/s	Assumptions Applied	Commissioned By/Collaboration With	Citation Record[a]
OECD/ITF (2016)	Lisbon municipality	SAV, DRS	Similar to above. No trip chaining or interaction of travel between household members. Maximum one transfer acceptable to transit users	Same as above	7
Martinez and Viegas (2017)	Lisbon municipality	SAV, DRS	Similar to above	Acknowledges ITF Corporate Partnership Board	47
OECD/ITF (2017)	Lisbon Metropolitan Area	SAV, DRS	Complex mode selection and transfer rules, assumed DRS travel speed, assumed costs	International Transport Forum's Corporate Partnership Board (CPB); involving: Brisa, Ford, Google, INRIX, PTV Group, SNCF, Transdev, Toyota, Uber	2
OECD/ITF (2018)	Small area (António Augusto Aguiar) in Lisbon (high intensity employment and retail plus affluent residential and high density bus boardings)	DRS	Pick-up behaviours	International Transport Forum's Corporate Partnership Board (CPB), involving: Abertis, Alstom, Brisa, Ford Motor Company, Kapsch TrafficCom, PTV Group, RATP, Renault-Nissan Alliance, Siemens, SNCF, Toyota Motor Corporation, Uber, Volvo Car Corporation. Draws on ITF CPB Workshop with participants from: Michelin, PTV-Group, Keolis, Toyota, RATP, Alstom, TransDev, Uber, Abertis, ITF	No data

Miscellaneous Other Studies					
Burghout, Rigole, and Andreasson (2015)	Stockholm	SAV, DRS	Replace all work PCV trips. Link speed based on free flow adjusted down, ride-sharing rules	Supported by Transportekonomiska forskningsstiftelsen (Transport Economics Research Foundation); demand data and guidance from KTH Transport and Location Analysis	43
Bischoff and Maciejewski (2016)	Berlin	SAV	Replacing all private car trips only, no external trips, fixed time increment link speeds	Supported by the Einstein Foundation Berlin, Germany, and the National Centre for Research and Development, Poland (grant ERA-NET-TRANSPORT-III/2/2014)	100
Boesch, Ciari, and Axhausen (2016)	Zurich region (422 by 272 km)	SAV	Assumed 5-10 minute waiting time, assume only existing car trips would shift to AVs, use by random sample of travellers	Draws on work developed at the Institut für Verkehrsplanung und Transportsysteme, Eidgenössische Technische Hochschule Zürich	91
Heilig, Hilgert, Mallig, Kagerbauer, and Vortisch (2017)	Stuttgart Region	DRS	Replace all PCVs with DRS – assume up to 4 ppl/ride	Developed at the Profilregion Mobilitätssysteme Karlsruhe, which is funded by the Ministry of Economic Affairs, Labour and Housing in Baden-Württemberg and as a national High Performance Center by the Fraunhofer-Gesellschaft	8
Meyer, Becker, Bösch, and Axhausen (2017)	Switzerland (entire country)	PAV, SAV	Existing model excludes intra-zonal and intra-urban demand. Link speed correlated to volume	Uses Swiss national transport model. Funded within the Swiss National Science Foundation's project 'Autonomous Cars - The next revolution in mobility' (grant number 200021_159234)	88

Table 2. (*Continued*)

Reference	Model Geography	Vehicle Type/s	Assumptions Applied	Commissioned By/Collaboration With	Citation Record[a]
Jäger, Agua, and Lienkamp (2017)	Munich (subsets)	SAV (all EVs)	Assumes all car demand is served by SAVs. Fixed traffic factors	Not stated	4
Australia					
KPMG Melbourne Group					
Thakur, Kinghorn, and Grace (2016)	Melbourne	PAV, SAV	VOTT, costs, fixed annual vehicle distance travelled, mean trip length and travel speed, fleet size	Not stated, uses KPMI LUTI model	11
KPMG (2018)	Melbourne	PAV, SAV 4&5 EVs	Mode Share, Land use, VOTT, Vehicle types, Fleet Size, Link flow, pricing structures. Assume utilisation of public transport is directly related to road congestion	KPMG commissioned by Infrastructure Victoria, using the Melbourne Activity-Based Model	No data
Brisbane (Davidson and Spinoulas) Group					
Davidson and Spinoulas (2015)	Brisbane, Queensland	PAV, SAV, DRS	Stochastically varied cost, VOTT, utility values, trip rate multiples, capacity	Not stated; uses own proprietary model	27
Davidson and Spinoulas (2016)	South East Queensland	PAV, SAV, DRS	Uniform adoption rate of AVs across model geography	Notes work undertaken for Queensland Department of Transport and Main Roads. Uses own proprietary model	20

Asia					
Singapore (various research groups)					
Spieser et al. (2014)	Singapore	PAV, SAV	Replacing all 'personal transport' modes with SAVs Derived link speed	Partially supported by the Singapore National Research Foundation, under the Future Urban Mobility SMART IRG Program. Household travel survey data provided by the Land Transport Authority of Singapore	275
Azevedo et al. (2016)	Subset of Singapore's car-restricted area	SAV	SAV operating only in car restricted zone. Assumed SAV is 40% of cost of conventional taxi. Parking not considered. No resident private vehicle use allowed. Assumed driving performance parameters	Supported by the National Research Foundation Singapore through the Singapore MIT Alliance for Research and Technology; data supplied by the Land Transport Authority and the Urban Redevelopment Authority	41
Shen, Zhang, and Zhao (2018)	Tampines Station catchment, Singapore	SAV and DRS	Lowest demand bus routes to station are modified. The remaining 10% of demand serviced by those routes is served by SAV/DRS. Maximum 10 min wait time. Distance based pricing	Supported by the National Research Foundation (NRF), Prime Minister's Office, Singapore, under CREATE programme, Singapore-MIT Alliance for Research and Technology (SMART) Centre, Future Urban Mobility (FM) IRG	27

[a]Data from Google Scholar, as at October 2019.

In our experience, the 'Lisbon' (OECD) and 'Boston' studies are particularly influential in practice circles. An AV 'end state' focussing on Level 5 (fully autonomous) driving was the main focus of models rather than providing intelligence on AVs emerging within the existing road user mix. This is a shortcoming since the emerging mobility technologies will be part of a mixed transport offering for many years. Modelling often combined forecast benefits of mobility types, particularly where both autonomous and electric vehicle (EV) technologies were assumed as available.

Smart Mobility 'Knowledge'

Taken as a whole, the suite of modelling studies we reviewed were heavily focussed on transport impacts rather than urban form impacts. Some impacts are consistent across studies, but many findings are mixed depending on input data and assumptions and model geography. Consistent messages are that, by mode, PAVs present the greatest negative impact whereas SAV had less impact – but this typically depended on whether the modeller assumed that a backbone public transport system would be complemented or competed with by SAVs. The DRS mode presented improved outcomes (but still negatives exist), but its take up will depend on the propensity of PCV drivers to share.

Limitations of these PSTs to Policy-makers

Critical to any understanding of knowledge drawn from these studies is the need to examine the assumptions made and the data used. Papers differed significantly in their description of assumptions: many did not describe a full set of assumptions – for example, assumptions about the existence and competitiveness of public transport were described in some papers, but not in others.

Data, Inputs, and Model Design

As models apply normative assumptions and value judgements to input data, the quality and relevance of base data is critical, particularly if findings are later applied by others to other cases. The general impacts of AVs have been hypothesised across a wide set of literature (Clements & Kockelman, 2017; Milakis, van Arem, & van Wee, 2017), and these lines of thinking are often translated into model assumptions, which then correlate with model outputs (Soteropoulos, Berger, & Ciari, 2019). The reviewed papers appear to reflect a line of inquiry which extrapolates (rather than evaluates or challenges) a very limited set of preliminary and speculative research. A central challenge for modellers faced with an unknown future is how to re-evaluate traditional assumptions and judgements on the nature of transport demand. This step is a critical given the instrumental role of modelling outputs in policy development, and/or project analysis. It is also critical that these assumptions are made explicit.

The data and inputs used to generate trip demand are critical to transport modelling (Ortúzar & Willumsen, 2011). Most of studies used either generated synthetic trip demand data based on historic travel behaviour surveys (mostly from around 2005–2012), or pre-existing demand forecasts for cities generated by conventional transport modelling as the primary input. This poses several potential sources of error. The age, sample sizes, accuracy and potential biases of these surveys notwithstanding (see Millard-Ball, 2015), past travel demands have been captured in a context in which AVs have not existed. Such data are inherently shaped both by the modal and spatial accessibility options available at the time. Existing transport models have been developed over a period of relatively stable urban transport conditions (Hawkins & Nurul Habib, 2018, p. 13), and may be flawed in the assessment of disruptive transformations in future mobility – thus limiting the degree to which past observations can support future predictions (Marsden & McDonald, 2019).

Many papers applied assumptions about the desirability of AV modes, such as using varied Value of Travel Time (VOTT) values or acceptable maximum waiting time parameters (see Singleton, 2018 for a review and critique). While understandable, early stated preference survey data are available (Anania, Rice, Walters, et al., 2018; Payre, Cestac, & Delhomme, 2014), and research for analogous modes, such as car-sharing (Firnkorn & Müller, 2015; Kent, Dowling, & Maalsen, 2017) and ride-sourcing (Rayle, Dai, Chan, Cervero, & Shaheen, 2016), enables improvement to these assumptions. Some evidence suggests that AV in-vehicle travel time will be no more useful than train travel (Noruzoliaee, Zou, & Liu, 2018).

Research groups are now broadening their work to include extensive original surveying specific to AVs (see Bansal and Kockelman, 2017). While this is may improve accuracy, it will be important to identify the geographic and demographic variability of critical preferences and likely behavioural patterns within their socio-economic context if research findings are to form inputs to decision-making in other places. Sensitivity-testing approaches do reveal interesting alternative trajectories (see Childress et al., 2015, who test the effects of parameters such as VOTT).

Trip Purpose and Demography

A large proportion of studies generated trips using activity-based methods, but lacked segmentation by trip purpose or demographic group. Demand was treated as homogenous, or modelling explored only employment commuting. This is concerning given the emphasis on equity considered fundamental to public interest transport planning (Pereira, Schwanen, & Banister, 2017). Where demographics were integrated into trip forecasting, they appeared to do so at a basic level (categorising 'work' and 'non-work' trips), which does not account for the complexity of personal travel routines. This shortfall will render results less useful, since utilisation of AV services is likely to be vary by trip type. Intercept surveying of ride-sharing customers in San Francisco (Rayle et al., 2016) suggests that social and leisure journey purposes represent a very large (>50%) share of all ridesharing trips; and mobility services are particularly desirable for alcohol-impaired trips (Payre et al., 2014). Factors which underpin trip-making behaviours for these activities may not be fully accounted within conventional trip valuation practices.

Trip purpose and demography are critical in understanding sources of latent demand (Clifton & Moura, 2017). Selection and market penetration of the various modes of AVs is likely to vary by these factors (Choudhury, Yang, de Abreu e Silva, & Ben-Akiva, 2017), particularly when trips are multimodal (Yap, Correia, & van Arem, 2016). Estimates of potentially new and coordinated travel by people who are presently not able to operate a motor vehicle vary (Harper, Hendrickson, Mangones, & Samaras, 2016; Kröger, Kuhnimhof, & Trommer, 2018). Comparisons are made more difficult due to the metric used (trips, VKT, etc.) – for example, Truong, De Gruyter, Currie, and Delbosc (2017) forecast 4% increase in total daily trips in Victoria, Australia by the very young or old. Kröger et al. (2018) predict that between 65% and 88% of new VMT triggered by AVs in groups (not disaggregated) who are not able to drive. Patterns of travel among children and older people are complex, heavily influenced by perceptions of personal safety, amenity, by both the individual, and their family or caregivers. Low driving distances among older people may not translate to greater utilisation of AVs if fear of travel is of personal safety rather than lack of confidence in driving. Parent's willingness to let children ride independently will significantly influence use of AV services by younger people (Anania, Rice, Winter et al., 2018); the early knowledge base for this remains limited. This may be further tempered by commercial operators' willingness to accept children as independent travellers.

Some studies only modelled AVs for a specific application, particularly for first/last mile trips to and from transit (e.g. Shen et al., 2018). This approach, especially when paired with localised surveying (e.g. Liang et al., 2016), is likely to yield very specific insights of direct use for transit planning agencies, but may miss broader mode shifts that might result from AV proliferation.

Geographic Variation

Studies concentrate in a small number of developed cities, particularly in Western Europe, the United States, and Singapore. Cities, such as Austin, Texas, and Sioux Falls, South Dakota, have received greatest attention. The over-representation of studies in cities or regions with low density, car-oriented land use structures and relatively uncompetitive public transport systems limit the representativeness of those results for policy-making in other cities. Studies in denser cities with more mature public transport networks (Singapore, Seoul, and Western European cites) also did not model, in detail, the manner in which trips may become more fragmented or integrated with public transport networks.

We did not identify a relevant study for any city in Africa, South America, or central Asia. Similarly, no study for a city in a developing country was identified. Further study of AVs in developing cities will be of value, particularly where informal taxi and bus services are common and analogous to SAV and DRS services.

Network Capacity Assumptions

Many papers assume AVs will alter the performance and capacity of existing road links (e.g. Childress et al., 2015). Some assumptions, such as that all speed limits will be increased (Davidson & Spinoulas, 2016), run directly counter to sustainable transport planning principles, such as reducing arterial road travel speeds to prevent road trauma and improve utilisation of active and public transport. These performance-oriented assumptions are likely to neglect other (more sustainable) modes; for instance, Patel et al. (2016) note that automated intersection controls with reduced headways may be 'disconcerting' to human drivers, without reference to the potential reactions of pedestrians. For the majority of streets within cities, the expected traffic flow optimisations of AVs may be negligible because the mix of road users and conflicting street uses will largely constrain AVs to behave cautiously in a manner similar to current driving behaviour. Some researchers (e.g. Millard-Ball, 2018) have suggested that AVs might be expected to perform perfectly in yielding to pedestrians, reframing the relationship with road users to an ultimately safer (and slower) system of mobility within cities. This will depend on regulation relating to self-driving entities, which remain contentious and uncertain.

Modal Shares and Impact on Public Transport

A potential source of overestimation in AV modelling arises where shifts to other modes are not considered (Heilig et al., 2017). Many sources simply assume complete replacement of PCVs with some form of AVs (Truong et al., 2017), or apply simplistic rules to assess active transport modes. Full proliferation and associated economies of scale is often assumed, but this not be realised until the conclusion of a long transition period (Merlin, 2017). Such assumptions about complete transfers of demand between mode, even for categories of trips, may neglect significant variation and diversity in travel demand, especially in different urban contexts. AVs may replace some forms of public transport, while complementing others, depending on context (McLeod, Scheurer, & Curtis, 2017). Many studies similarly applied basic assumptions about AV costs and pricing (Azevedo et al., 2016; Hörl et al., 2016; Thakur et al., 2016; Zhao & Kockelman, 2018), which may not represent the dynamic pricing of different ridesharing, MaaS and AV services.

The lack of revealed preference data for use of AVs hinders effective modelling (Hawkins & Nurul Habib, 2018). Yet there is a growing body of AV mode choice and competitiveness survey data now available that enable more granular, disaggregated, and conditional analysis of mode shift potential (Becker & Axhausen, 2017; Bösch, Becker, Becker, & Axhausen, 2018; Chen, Zahiri, & Zhang, 2017; Haboucha, Ishaq, & Shiftan, 2017; Lu, Du, Dunham-Jones, Park, & Crittenden, 2017). As the pricing of SAV and DRS services is likely to be dynamic depending on demand, greater sensitivity testing of pricing options may be of value for policy-makers.

Transport Demand, Location, and Induced Demand

Studies either ignored or over-simplified the complex interdependences between spatial accessibility, land use distribution, economic activity, competing citizen preferences, and transport demand. Studies commonly did not attempt to model the effect of AV service on travel behaviour decisions (see Masoud & Jayakrishnan, 2017), such as in destination choices (e.g. Liu et al., 2017). Early Lisbon studies (Martinez, Correia, & Viegas, 2015; OECD/ITF, 2015) appear to assume all trip demand is realised, despite evidence that people generally adjust their travel behaviour according to congestion and other impeding factors.

The degree to which AVs may alter urban structure and the relationships which underpin travel patterns in cities is a key concern to planners, and has received preliminary modelling focus (Hawkins & Nurul Habib, 2018; KPMG, 2018), though it was not a common feature. These interdependencies between access, land use distribution, human activity, and transport patterns are not passive or coincidental. They drive, and are shaped by, processes and decisions of city and regional planning.

The Roles for Other PSTs in Shaping 'Smart' Urban Futures

Jurisdictions are concerned with the potential impact of AVs and how to plan and govern for them. While the use of transport models as a PST remains commonplace, though controversial, there are fundamental problems in interpreting much of this highly specialised technical research for practical use in policy-making. Comparing AV modelling research is challenging, due to the wide diversity of modelling platforms and tools used, across a wide range of hypothetical spatial and non-spatial analyses. Nevertheless, it is critical that the mechanism, findings, and implications of such studies are evaluated.

Transport demand models and their underlying inputs are often designed to represent the highest possible demand, to enforce generous design and perceived 'future-proofing' of infrastructure (Millard-Ball, 2015), with limited regard for other implications. They have been a primary instrument in which substantive theories of road planning have been transplanted across contexts, with little regard for local context and conditions (Joutsiniemi & Curtis, 2015). The planning 'lock-in' created through self-fulfilling demand forecasts which then become induced by projects; and a propensity for models to be manipulated for the benefit of project proponents (Curtis & Low, 2012; Kenworthy, 2012; Næss, 2006) adds a further concern.

As models become more complex, their usefulness can diminish in practical decision-making, principally because models become harder for stakeholders understand and apply in decision-making (Lee, 1973; Saujot et al., 2016; te Brömmelstroet, 2010), as well as more cumbersome to run. There is an inherent tendency for models to reflect normative biases, existing preconceptions and implicit values. Yet modelling studies have played a role in planning strategy formation, needs analysis, and project justification. A key issue is where modelling studies should sit in the planning process. Should they be instrumental to policy – so leading long range planning decisions, or are they more usefully utilised as one input into the review of options for future urban development or projects that form part of the strategic plan? These core questions deserve further discussion. The fundamental idea of 'policy' is that it serves to guide decision-making. In a future mobility context such policy might relate to desired urban structure, investment in future transport systems, and regulation necessary to protect public interests. There is a risk that the outputs of modelling studies serve to lead policy-making, rather than as an instrument that serves to test, enact, or deliver policy based on a clear articulation of public interest objectives. The extent to which model outputs then guides policy development is an important unknown. A critical risk is that strategic transport policy might be developed based solely on the outputs of transport models, given the limitations we have noted.

Where transport models are used to test the impacts of a project, it is critical that decisions that follow are made in the context of the planning strategy. Development of a planning strategy is generally legitimised through democratic process – at the very least the public are consulted on its content. Where modelling activity occurs without reference to the planning strategy, policy can be undermined (Legacy, Curtis, & Scheurer, 2017). Where transport models project 'smart' mobility outcomes based on evidence drawn from past conditions of PCV-reliant urban structures and transport networks, decisions may perpetuate existing patterns and issues instead of capturing opportunities to realise transformative benefits in accessibility, amenity, and sustainability.

To avoid repeating past trajectories, strategic planning requires directive policy, remaining cognizant of key questions of public interest objectives. Given the nature of the potential transformations of driverless technology, such questions include: What should the vision for our future cities be? How can we capitalise on the benefits of smart mobility while avoiding the possible disbenefits? How do specific transport behaviours and patterns respond to changes in modes, service typologies, pricing, and regulation? This suggests the need to employ a wider suite of PSTs capable of integrating different forms of input information. PSTs must be capable of addressing the multi-dimensional complexity of people's daily activity needs, and the range of transport services (and non-transport policy options) that may serve them.

There is a role for a diverse range of other PSTs to serve as policy instruments. Use of PSTs in participatory contexts see the collective sharing, questioning, and building of knowledge about the future of cities. This approach can yield significant benefits for modellers, planners, and the wide range of stakeholders who should be part of defining objectives and doing practical planning (Pelzer & Geertman, 2014; te Brömmelstroet, Curtis, Larsson, & Milakis, 2016). An array of PSTs that have been used in practice offer value, such as:

- 'Map mash-up' workshops provide opportunities for high-level practitioners to explore urban futures, embracing issues of urban governance and complexity in an environment outside their usual practice. They bring together a diverse set of stakeholders to understand complex contextual dimensions of futures planning (Curtis, Scheurer, & McLeod, 2017).
- 'Backcasting' techniques enable stakeholders to role play and examine policy options again given targets (Hickman, Ashiru, & Banister, 2009).
- Participatory workshops provide a valuable environment in which to explore issues, policy options, and develop agreement on future directions, employing visualisation techniques and spatial mapping (Hartz-Karp, 2005; te Brömmelstroet, Silva, & Bertolini, 2014).

As they are typically used deliberatively, these PSTs engage with stakeholders in a more democratic way than conventional models.[2] A key feature is their ability to be used in involving stakeholders in open and transparent discussion. They are useful where the future is uncertain – supporting community learning, identifying institutional barriers or gaps in knowledge, aligning interests, charting implementation tactics, and coalescing political capital for policy change.

Conclusions

Cities are facing a succession of challenges posed by the emergence of new mobility technologies. Traditional modelling studies provide an insufficient knowledge base to inform policy for future mobility. For the most part, modelling has focussed on the new world cities which are presently structured to serve PCVs. They reflect normative assumptions about travel behaviour and preferences (Soteropoulos et al., 2019), draw on existing travel data based on simplistic representations of aggregated transport demand where PCVs dominate and there is a relative paucity of public transport options. They reveal little about the potential of new mobility forms on journey purpose or the attributes of potential AV users. Studies in Western Europe, Asia, and Australia have generally been limited to small case studies, and there is not a clear elucidation of how modelling tools have been applied within planning processes.

The lack of insight by trip purpose of demographic group reveals an implicit emphasis on commuting to work, which may constrain thinking. Some input data derived from public opinion and willingness to use emergent AV products are emerging, but its application may increase model complexity while simultaneously reducing model usability. Hence, greater focus on the purpose, methods, and outcomes of applying such models to assist in planning for future mobility will be of value to a wide set of researchers, modellers, practitioners, and policy-makers.

The reporting of models – particularly in the ambiguity of model assumptions and limitations – is likely to make interpretation for decision-makers challenging. Models differ from other PSTs in that they are intended to provide value through their outputs, which respond to a narrowly pre-conceived and precisely defined set of planning objectives, with decision-makers and stakeholders being largely insulated from the actual modelling process (Saujot et al., 2016). On the other hand, other PSTs are conceived to support processes of collaborative planning, where objectives may be explored and questioned, with the value of the tool realised during such evaluative discussions between a broader set of planning actors and stakeholders (Pelzer, Geertman, Heijden, & Rouwette, 2014; te Brömmelstroet et al., 2016).

There has been recurring and rigorous debate about the processes, reliability, and applications of conventional transport modelling (Atkins, 1987; Lee, 1973;

[2]Although we note Kronsell and Mukhtar-Landgren (2020, Chapter 7 in this volume) raise questions in this respect.

te Brömmelstroet, 2010). Optimism about AVs borne out of recent modelling research is therefore deserving of specific scrutiny, especially as the introduction of such technology will afford policy-makers with a unique transition step in which new norms about accessibility in cities may be defined (Legacy, Ashmore, Scheurer, Stone, & Curtis, 2018). Ultimately, transport models are one of many forms of 'knowledge technology' in planning (Gudmundsson, 2011). There remain several fundamental gaps in knowledge about future mobility which should temper the degree to which existing knowledge and assumptions are extrapolated. Fundamentally, we face an unknown future with an immense range of potentially interdependent changes (Milakis et al., 2017). Given the wide domains of knowledge involved, and the variation between research contexts, we urge caution. Future research must be interdisciplinary and draw on a range of analysis, where AV modelling provides one of many inputs used to inform decision-making.

References

Anania, E., Rice, S., Walters, N. W., Pierce, M., Winter, S. R., & Milner, M. N. (2018). The effects of positive and negative information on consumers' willingness to ride in a driverless vehicle. *Transport Policy*, *72*, 218–224. https://doi.org/10.1016/j.tranpol.2018.04.002

Anania, E., Rice, S., Winter, S., Milner, M., Walters, N., & Pierce, M. (2018). Why people are not willing to let their children ride in driverless school buses: A gender and nationality comparison. *Social Sciences*, *7*(3), 34. Retrieved from http://www.mdpi.com/2076-0760/7/3/34

Atkins, S. T. (1987). The crisis for transportation planning modelling. *Transport Reviews*, *7*(4), 307–325. doi:10.1080/01441648708716664

Auld, J., Sokolov, V., & Stephens, T. S. (2017). Analysis of the effects of connected–automated vehicle technologies on travel demand. *Transportation Research Record: Journal of the Transportation Research Board*, *2625*, 1–8. doi:10.3141/2625-01

Auld, J., Verbas, O., Javanmardi, M., & Rousseau, A. (2018). Impact of privately-owned level 4 CAV technologies on travel demand and energy. *Procedia Computer Science*, *130*(C), 914–919.

Azevedo, C. L., Marczuk, K., Raveau, S., Soh, H., Adnan, M., Basak, K., … Ben-Akiva, M. (2016). Microsimulation of demand and supply of autonomous mobility on demand. *Transportation Research Record: Journal of the Transportation Research Board*, *2564*, 21–30. doi:10.3141/2564-03

Bansal, P., & Kockelman, K. M. (2017). Forecasting Americans' long-term adoption of connected and autonomous vehicle technologies. *Transportation Research Part A: Policy and Practice*, *95*, 49–63. https://doi.org/10.1016/j.tra.2016.10.013

Becker, F., & Axhausen, K. W. (2017). Literature review on surveys investigating the acceptance of automated vehicles. *Transportation*, *44*(6), 1293–1306. doi:10.1007/s11116-017-9808-9

Bischoff, J., & Maciejewski, M. (2016). Simulation of city-wide replacement of private cars with autonomous taxis in Berlin. *Procedia Computer Science*, *83*, 237–244. https://doi.org/10.1016/j.procs.2016.04.121

Boesch, P. M., Ciari, F., & Axhausen, K. W. (2016). Autonomous vehicle fleet sizes required to serve different levels of demand. *Transportation Research Record: Journal of the Transportation Research Board*, *2542*, 111–119. doi:10.3141/2542-13

Bösch, P. M., Becker, F., Becker, H., & Axhausen, K. W. (2018). Cost-based analysis of autonomous mobility services. *Transport Policy*, *64*, 76–91. https://doi.org/10.1016/j.tranpol.2017.09.005

Brownell, C., & Kornhauser, A. (2014). A driverless alternative. *Transportation Research Record: Journal of the Transportation Research Board*, *2416*, 73–81. doi:10.3141/2416-09

Burghout, W., Rigole, P. J., & Andreasson, I. (2015). Impacts of shared autonomous taxis in a metropolitan area. Paper presented at the Proceedings of the 94th annual meeting of the Transportation Research Board, Washington DC.

Chen, T. D., & Kockelman, K. M. (2016). Management of a shared autonomous electric vehicle fleet. *Transportation Research Record: Journal of the Transportation Research Board*, *2572*, 37–46. doi:10.3141/2572-05

Chen, T. D., Kockelman, K. M., & Hanna, J. P. (2016). Operations of a shared, autonomous, electric vehicle fleet: Implications of vehicle & charging infrastructure decisions. *Transportation Research Part A: Policy and Practice*, *94*, 243–254. https://doi.org/10.1016/j.tra.2016.08.020

Chen, X., Zahiri, M., & Zhang, S. (2017). Understanding ridesplitting behavior of on-demand ride services: An ensemble learning approach. *Transportation Research Part C: Emerging Technologies*, *76*, 51–70. https://doi.org/10.1016/j.trc.2016.12.018

Childress, S., Nichols, B., Charlton, B., & Coe, S. (2015). Using an activity-based model to explore the potential impacts of automated vehicles. *Transportation Research Record: Journal of the Transportation Research Board*, *2493*, 99–106. doi:10.3141/2493-11

Choudhury, C. F., Yang, L., de Abreu e Silva, J., & Ben-Akiva, M. (2017). Modelling preferences for smart modes and services: A case study in Lisbon. *Transportation Research Part A: Policy and Practice*, *115*, 15–31. https://doi.org/10.1016/j.tra.2017.07.005

Clements, L. M., & Kockelman, K. M. (2017). Economic effects of automated vehicles. *Transportation Research Record: Journal of the Transportation Research Board*, *2606*, 106–114. doi:10.3141/2606-14

Clifton, K. J., & Moura, F. (2017). Conceptual framework for understanding latent demand: Accounting for unrealized activities and travel. *Transportation Research Record: Journal of the Transportation Research Board*, *2668*, 78–83.

Correia, G. H. d. A., & van Arem, B. (2016). Solving the user optimum privately owned automated vehicles assignment problem (UO-POAVAP): A model to explore the impacts of self-driving vehicles on urban mobility. *Transportation Research Part B: Methodological*, *87*, 64–88. https://doi.org/10.1016/j.trb.2016.03.002

Curtis, C., & Low, N. (2012). *Institutional barriers to sustainable transport*. Surrey: Ashgate.

Curtis, C., Scheurer, J., & McLeod, S. (2017). Map Mashups revisited: Employing discursive planning tools for Perth's urban future. Paper presented at the 8th State of Australian Cities National Conference (SOAC), Adelaide. Retrieved from http://apo.org.au/node/178871

Curtis, C., Stone, J., Legacy, C., & Ashmore, D. (2019). Governance of future urban mobility: A research agenda. *Urban Policy and Research*, *37*(3), 393–404. doi:10.1080/08111146.2019.1626711

Davidson, P., & Spinoulas, A. (2015). Autonomous vehicles: What could this mean for the future of transport. Paper presented at the Australian Institute of Traffic Planning and Management (AITPM) National Conference, Brisbane, Queensland.

Davidson, P., & Spinoulas, A. (2016). Driving alone versus riding together – How shared autonomous vehicles can change the way we drive. *Road & Transport Research*, *25*(3), 51–66. Retrieved from https://search.informit.com.au/documentSummary;dn=476853444665451

Evans, R., Burke, M., & Dodson, J. (2007). Clothing the emperor?: Transport modelling and decision-making in Australian cities. Paper presented at the Proceedings of State of Australian Cities National Conference 2007, Adelaide.

Fagnant, D. J., & Kockelman, K. M. (2014). The travel and environmental implications of shared autonomous vehicles, using agent-based model scenarios. *Transportation Research Part C: Emerging Technologies, 40*, 1–13. https://doi.org/10.1016/j.trc.2013.12.001

Fagnant, D. J., & Kockelman, K. M. (2018). Dynamic ride-sharing and fleet sizing for a system of shared autonomous vehicles in Austin, Texas. *Transportation, 45*(1), 143–158. doi:10.1007/s11116-016-9729-z

Fagnant, D. J., Kockelman, K. M., & Bansal, P. (2015). Operations of shared autonomous vehicle fleet for Austin, Texas, Market. *Transportation Research Record: Journal of the Transportation Research Board, 2536*, 98–106. doi:10.3141/2536-12

Firnkorn, J., & Müller, M. (2015). Free-floating electric carsharing-fleets in smart cities: The dawning of a post-private car era in urban environments? *Environmental Science & Policy, 45*, 30–40. https://doi.org/10.1016/j.envsci.2014.09.005

Gudmundsson, H. (2011). Analysing models as a knowledge technology in transport planning. *Transport Reviews, 31*(2), 145–159. doi:10.1080/01441647.2010.532884

Haboucha, C. J., Ishaq, R., & Shiftan, Y. (2017). User preferences regarding autonomous vehicles. *Transportation Research Part C: Emerging Technologies, 78*, 37–49. https://doi.org/10.1016/j.trc.2017.01.010

Harper, C. D., Hendrickson, C. T., Mangones, S., & Samaras, C. (2016). Estimating potential increases in travel with autonomous vehicles for the non-driving, elderly and people with travel-restrictive medical conditions. *Transportation Research Part C: Emerging Technologies, 72*, 1–9. https://doi.org/10.1016/j.trc.2016.09.003

Hartz-Karp, J. (2005). A case study in deliberative democracy: Dialogue with the city. *Journal of Public Deliberation, 1*(1), 6. Retrieved from http://www.publicdeliberation.net/jpd/vol1/iss1/art6/

Hawkins, J., & Nurul Habib, K. (2018). Integrated models of land use and transportation for the autonomous vehicle revolution. *Transport Reviews, 39*, 66–83. doi:10.1080/01441647.2018.1449033

Heilig, M., Hilgert, T., Mallig, N., Kagerbauer, M., & Vortisch, P. (2017). Potentials of autonomous vehicles in a changing private transportation system – A case study in the stuttgart region. *Transportation Research Procedia, 26*, 13–21. https://doi.org/10.1016/j.trpro.2017.07.004

Hickman, R., Ashiru, O., & Banister, D. (2009). Achieving carbon-efficient transportation: Backcasting from London. *Transportation Research Record: Journal of the Transportation Research Board, 2139*, 172–182. doi:10.3141/2139-20

Hörl, S., Erath, A., & Axhausen, K. W. (2016). Simulation of autonomous taxis in a multi-modal traffic scenario with dynamic demand. *Arbeitsberichte Verkehrs-und Raumplanung*, [*Working Paper Series: Transport and spatial planning*] *1184*.

Jäger, B., Agua, F. M. M., & Lienkamp, M. (2017,). Agent-based simulation of a shared, autonomous and electric on-demand mobility solution. Paper presented at the 2017 IEEE 20th International Conference on Intelligent Transportation Systems (ITSC), 16–19 October.

Joutsiniemi, A., & Curtis, C. (2015). Unlikely similar – Forking paths and path dependency in transportation studies of Helsinki and Melbourne. Paper presented at the AESOP Conference, 13–16 July, Prague.

Kent, J., Dowling, R., & Maalsen, S. (2017). Catalysts for transport transitions: Bridging the gap between disruptions and change. *Journal of Transport Geography, 60*, 200–207. https://doi.org/10.1016/j.jtrangeo.2017.03.013

Kenworthy, J. (2012). Don't shoot me, I'm only the transport planner (apologies to Sir Elton John). *World Transport Policy and Practice, 18*(4), 6–26.

KPMG. (2018). Automated and zero emission vehicle infrastrucuture advice – Transport modelling. Retrieved from https://www.infrastructurevictoria.com.au/project/automated-and-zero-emission-vehicle-infrastructure/

Kronsell, A., & Mukhtar-Landgren, D. (2020). Experimental governance of smart mobility. Some normative implications. In A. Paulsson & C. H. Sørensen (Eds.), *Shaping smart mobility futures – Governance and policy instruments in times of sustainability transitions* (pp. 119–135). Bingley: Emerald.

Kröger, L., Kuhnimhof, T., & Trommer, S. (2018). Does context matter? A comparative study modelling autonomous vehicle impact on travel behaviour for Germany and the USA. *Transportation Research Part A: Policy and Practice, 122*, 146–161. https://doi.org/10.1016/j.tra.2018.03.033

Lee, D. B. (1973). Requiem for large-scale models. *Journal of the American Institute of Planners, 39*(3), 163–178. doi:10.1080/01944367308977851

Legacy, C., Ashmore, D., Scheurer, J., Stone, J., & Curtis, C. (2018). Planning the driverless city. *Transport Reviews, 39*, 1–19. doi:10.1080/01441647.2018.1466835

Legacy, C., Curtis, C., & Scheurer, J. (2017). Planning transport infrastructure: Examining the politics of transport planning in Melbourne, Sydney and Perth. *Urban Policy and Research, 35*(1), 44–60. doi:10.1080/08111146.2016.1272448

Levin, M. W., & Boyles, S. D. (2015). Effects of autonomous vehicle ownership on trip, mode, and route choice. *Transportation Research Record: Journal of the Transportation Research Board, 2493*, 29–38. doi:10.3141/2493-04

Levin, M. W., Kockelman, K. M., Boyles, S. D., & Li, T. (2017). A general framework for modeling shared autonomous vehicles with dynamic network-loading and dynamic ride-sharing application. *Computers, Environment and Urban Systems, 64*, 373–383. https://doi.org/10.1016/j.compenvurbsys.2017.04.006

Liang, X., Correia, G. H. d. A., & van Arem, B. (2016). Optimizing the service area and trip selection of an electric automated taxi system used for the last mile of train trips. *Transportation Research Part E: Logistics and Transportation Review, 93*, 115–129. https://doi.org/10.1016/j.tre.2016.05.006

Liu, J., Kockelman, K. M., Boesch, P. M., & Ciari, F. (2017). Tracking a system of shared autonomous vehicles across the Austin, Texas network using agent-based simulation. *Transportation, 44*(6), 1261–1278. doi:10.1007/s11116-017-9811-1

Llorca, C., Moreno, A., & Moeckel, R. (2017). Effects of shared autonomous vehicles on the level of service in the greater Munich metropolitan area. Paper presented at the International Conference on Intelligent Transport Systems in Theory and Practice, mobil.TUM, Munich.

Loeb, B., Kockelman, K. M., & Liu, J. (2018). Shared autonomous electric vehicle (SAEV) operations across the Austin, Texas network with charging infrastructure decisions. *Transportation Research Part C: Emerging Technologies, 89*, 222–233. https://doi.org/10.1016/j.trc.2018.01.019

Lu, Z., Du, R., Dunham-Jones, E., Park, H., & Crittenden, J. (2017). Data-enabled public preferences inform integration of autonomous vehicles with transit-oriented development in Atlanta. *Cities, 63*, 118–127. https://doi.org/10.1016/j.cities.2017.01.004

Marsden, G., & McDonald, N. C. (2019). Institutional issues in planning for more uncertain futures. *Transportation, 46*(4), 1075–1092. doi:10.1007/s11116-017-9805-z

Marsden, G., & Reardon, L. (2017). Questions of governance: Rethinking the study of transportation policy. *Transportation Research Part A: Policy and Practice, 101*, 238–251. https://doi.org/10.1016/j.tra.2017.05.008

Martinez, L. M., Correia, G. H. d. A., & Viegas, J. M. (2015). An agent-based simulation model to assess the impacts of introducing a shared-taxi system: An application to Lisbon (Portugal). *Journal of Advanced Transportation, 49*(3), 475–495. doi:10.1002/atr.1283

Martinez, L. M., & Viegas, J. M. (2017). Assessing the impacts of deploying a shared self-driving urban mobility system: An agent-based model applied to the city of Lisbon, Portugal. *International Journal of Transportation Science and Technology, 6*(1), 13–27. https://doi.org/10.1016/j.ijtst.2017.05.005

Masoud, N., & Jayakrishnan, R. (2017). Autonomous or driver-less vehicles: Implementation strategies and operational concerns. *Transportation Research Part E: Logistics and Transportation Review, 108*, 179–194. https://doi.org/10.1016/j.tre.2017.10.011

McLeod, S., Scheurer, J., & Curtis, C. (2017). Urban public transport: Planning principles and emerging practice. *Journal of Planning Literature, 32*(3), 223–239. doi:10.1177/0885412217693570

Mendes, L. M., Bennàssar, M. R., & Chow, J. Y. J. (2017). Comparison of light rail streetcar against shared autonomous vehicle fleet for Brooklyn–Queens Connector in New York City. *Transportation Research Record: Journal of the Transportation Research Board, 2650*, 142–151. doi:10.3141/2650-17

Merlin, L. M. (2017). Comparing automated shared taxis and conventional bus transit for a small city. *Journal of Public Transportation, 20*(2), 19–39. doi:10.5038/2375-0901.20.2.2

Meyer, J., Becker, H., Bösch, P. M., & Axhausen, K. W. (2017). Autonomous vehicles: The next jump in accessibilities? *Research in Transportation Economics, 62*, 80–91. https://doi.org/10.1016/j.retrec.2017.03.005

Milakis, D., van Arem, B., & van Wee, B. (2017). Policy and society related implications of automated driving: A review of literature and directions for future research. *Journal of Intelligent Transportation Systems, 21*(4), 324–348. doi:10.1080/15472450.2017.1291351

Millard-Ball, A. (2015). Phantom trips: Overestimating the traffic impacts of new development. *8*(1), 19. doi:10.5198/jtlu.2015.384

Millard-Ball, A. (2018). Pedestrians, autonomous vehicles, and cities. *Journal of Planning Education and Research, 38*(1), 6–12. doi:10.1177/0739456x16675674

Moreno, A. T., Michalski, A., Llorca, C., & Moeckel, R. (2018). Shared autonomous vehicles effect on vehicle-km traveled and average trip duration. *Journal of Advanced Transportation, 2018*, 8969353. doi:10.1155/2018/8969353

Næss, P. (2006). Cost-benefit analyses of transportation investments. *Journal of Critical Realism, 5*(1), 32–60. doi:10.1558/jocr.v5i1.32

Noruzoliaee, M., Zou, B., & Liu, Y. (2018). Roads in transition: Integrated modeling of a manufacturer-traveler-infrastructure system in a mixed autonomous/human driving environment. *Transportation Research Part C: Emerging Technologies, 90*, 307–333. https://doi.org/10.1016/j.trc.2018.03.014

OECD/ITF. (2015). Urban mobility system upgrade: How shared self-driving cars could change city traffic. Retrieved from https://www.itf-oecd.org/sites/default/files/docs/15cpb_self-drivingcars.pdf

OECD/ITF. (2016). Shared mobility: Innovation for liveable cities. Retrieved from https://www.itf-oecd.org/sites/default/files/docs/shared-mobility-liveable-cities.pdf

OECD/ITF. (2017). Transition to shared mobility: How large cities can deliver inclusive transport services. Retrieved from https://www.itf-oecd.org/transition-shared-mobility

OECD/ITF. (2018). The shared-use city: Managing the curb, Paris. Retrieved from https://www.itf-oecd.org/shared-use-city-managing-curb-0

Ortúzar, J. d. D., & Willumsen, L. G. (2011). *Modelling transport*. Chichester: John Wiley & Sons.

Patel, R., Levin, M. W., & Boyles, S. D. (2016). Effects of autonomous vehicle behavior on arterial and freeway networks. *Transportation Research Record: Journal of the Transportation Research Board, 2561*, 9–17. doi:10.3141/2561-02

Payre, W., Cestac, J., & Delhomme, P. (2014). Intention to use a fully automated car: Attitudes and a priori acceptability. *Transportation Research Part F: Traffic Psychology and Behaviour, 27*, 252–263. https://doi.org/10.1016/j.trf.2014.04.009

Pelzer, P., & Geertman, S. (2014). Planning support systems and interdisciplinary learning. *Planning Theory & Practice, 15*(4), 527–542. doi:10.1080/14649357.2014.963653

Pelzer, P., Geertman, S., Heijden, R. v. d., & Rouwette, E. (2014). The added value of planning support systems: A practitioner's perspective. *Computers, Environment and Urban Systems, 48*, 16–27. doi:http://dx.doi.org/10.1016/j.compenvurbsys.2014.05.002

Pereira, R. H. M., Schwanen, T., & Banister, D. (2017). Distributive justice and equity in transportation. *Transport Reviews, 37*(2), 170–191. doi:10.1080/01441647.2016.1257660

Rayle, L., Dai, D., Chan, N., Cervero, R., & Shaheen, S. (2016). Just a better taxi? A survey-based comparison of taxis, transit, and ridesourcing services in San Francisco. *Transport Policy, 45*, 168–178. doi:http://dx.doi.org/10.1016/j.tranpol.2015.10.004

Saujot, M., de Lapparent, M., Arnaud, E., & Prados, E. (2016). Making land use – Transport models operational tools for planning: From a top-down to an end-user approach. *Transport Policy, 49*, 20–29. doi:http://dx.doi.org/10.1016/j.tranpol.2016.03.005

Shen, W., & Lopes, C. (2015). Managing autonomous mobility on demand systems for better passenger experience. Paper presented at the PRIMA 2015: Principles and Practice of Multi-Agent Systems, 26–30 October, Bertinoro, Italy.

Shen, Y., Zhang, H., & Zhao, J. (2018). Integrating shared autonomous vehicle in public transportation system: A supply-side simulation of the first-mile service in Singapore. *Transportation Research Part A: Policy and Practice, 113*, 125–136. https://doi.org/10.1016/j.tra.2018.04.004

Singleton, P. A. (2018). Discussing the "positive utilities" of autonomous vehicles: Will travellers really use their time productively? *Transport Reviews, 39*, 50–65. doi:10.1080/01441647.2018.1470584

Soteropoulos, A., Berger, M., & Ciari, F. (2019). Impacts of automated vehicles on travel behaviour and land use: an international review of modelling studies. *Transport Reviews, 39*(1), 29–49. doi:10.1080/01441647.2018.1523253

Spieser, K., Treleaven, K., Zhang, R., Frazzoli, E., Morton, D., & Pavone, M. (2014). Toward a systematic approach to the design and evaluation of automated mobility-on-demand systems: A case study in Singapore. In G. Meyer & S. Beiker (Eds.), *Road vehicle automation* (pp. 229–245). Cham: Springer International Publishing.

te Brömmelstroet, M. (2010). Equip the warrior instead of manning the equipment: Land use and transport planning support in the Netherlands. *The Journal of Transport and Land Use, 3*(1), 25–41. doi:10.5198/jtlu.v3i1.99

te Brömmelstroet, M., Curtis, C., Larsson, A., & Milakis, D. (2016). Strengths and weaknesses of accessibility instruments in planning practice: Technological rules based on experiential workshops. *European Planning Studies, 24*, 1175–1196. doi:10.1080/09654313.2015.1135231

te Brömmelstroet, M., Silva, C., & Bertolini, L. (2014). Assessing usability of accessibility instruments. Retrieved from http://www.accessibilityplanning.eu/wp-content/uploads/2014/05/COST-REPORT-II.pdf

Thakur, P., Kinghorn, R., & Grace, R. (2016). Urban form and function in the autonomous era. Paper presented at the Australasion Transport Research Forum (ATRF), 16–18 November, Melbourne.

The Boston Consulting Group. (2017). Making autonomous vehicles a reality: Lessons from Boston and beyond. Retrieved from https://www.bcg.com/en-au/publications/2017/automotive-making-autonomous-vehicles-a-reality.aspx

Truong, L. T., De Gruyter, C., Currie, G., & Delbosc, A. (2017). Estimating the trip generation impacts of autonomous vehicles on car travel in Victoria, Australia. *Transportation, 44*(6), 1279–1292. doi:10.1007/s11116-017-9802-2

Van Wee, B., & Banister, D. (2016). How to write a literature review paper? *Transport Reviews, 36*(2), 278–288. doi:10.1080/01441647.2015.1065456

Winter, K., Cats, O., Correia, G. H. d. A., & Arem, B. v. (2016). Designing an automated demand-responsive transport system. *Transportation Research Record: Journal of the Transportation Research Board, 2542*, 75–83. doi:10.3141/2542-09

World Economic Forum. (2018). Reshaping urban mobility with autonomous vehicles lessons from the city of Boston. Retrieved from https://www.weforum.org/reports/reshaping-urban-mobility-with-autonomous-vehicles-lessons-from-the-city-of-boston

Yap, M. D., Correia, G. H. d. A., & van Arem, B. (2016). Preferences of travellers for using automated vehicles as last mile public transport of multimodal train trips. *Transportation Research Part A: Policy and Practice*, *94*, 1–6. https://doi.org/10.1016/j.tra.2016.09.003

Zachariah, J., Gao, J., Kornhauser, A., & Mufti, T. (2014). Uncongested mobility for all: A proposal for an area wide autonomous taxi system in New Jersey. Paper presented at the Transportation Research Board 93rd Annual Meeting, Washington DC. Retrieved from https://trid.trb.org/view/1288288

Zhang, W., & Guhathakurta, S. (2017). Parking spaces in the age of shared autonomous vehicles. *Transportation Research Record: Journal of the Transportation Research Board*, *2651*, 80–91. doi:10.3141/2651-09

Zhang, W., Guhathakurta, S., Fang, J., & Zhang, G. (2015). Exploring the impact of shared autonomous vehicles on urban parking demand: An agent-based simulation approach. *Sustainable Cities and Society*, *19*, 34–45. https://doi.org/10.1016/j.scs.2015.07.006

Zhang, W., Guhathakurta, S., & Khalil, E. B. (2018). The impact of private autonomous vehicles on vehicle ownership and unoccupied VMT generation. *Transportation Research Part C: Emerging Technologies*, *90*, 156–165. https://doi.org/10.1016/j.trc.2018.03.005

Zhao, Y., & Kockelman, K. M. (2018). Anticipating the regional impacts of connected and automated vehicle travel in Austin, Texas. *Journal of Urban Planning and Development*, *144*(4), 04018032. doi:10.1061/(ASCE)UP.1943-5444.0000463

Chapter 6

Challenges for Government as Facilitator and Umpire of Innovation in Urban Transport: The View from Australia

John Stone, David Ashmore, Crystal Legacy and Carey Curtis

ABSTRACT

New economies based on emerging technologies for shared mobility and autonomous vehicles will shape future urban transport systems, but their potential impacts are uncertain. Internationally, government agencies face difficult challenges to effectively plan and regulate the deployment of these technologies for the common good, whilst simultaneously encouraging innovation. Being both a facilitator and an umpire is not an easy task. This chapter draws on a series of interviews with public and private-sector actors in urban transport in Australia. Unsurprisingly, all private-sector respondents had significant concerns for the sustainability of their business in the emerging mobility markets, but it was generally acknowledged that without government support and partnership, a lack of structure and clarity could lead to natural monopolies with negative consequences for competition and the public good. Strong and clear government regulation is seen to be necessary to allow the sector to reach its maximum potential and have positive ramifications for both the public and the private good – outcome not always seen as compatible. Public-sector interviewees generally recognised that much of the necessary innovation was being shaped by the market, and that there had been a considerable loss of skills over decades from the state because of neo-liberal policies. So, some doubted the ability of the state to shape developments using currently available planning and public policy methods and feared that it would be difficult to regulate emergent markets to prevent monopolies emerging. On the other hand, some argued that many firms are looking to government for frameworks in which businesses can operate successfully by setting conditions in which risks could be managed.

Shaping Smart Mobility Futures:
Governance and Policy Instruments in times of Sustainability Transitions, 105–118

doi:10.1108/978-1-83982-650-420201006

This chapter discusses these issues, seeking to guide research agendas and to foster further debate. The evidence gained from these in-depth interviews helps focus attention on which forms of regulation might be required by industry. It also raises questions about the capacity of government agencies to effectively manage these complex transitions.

Keywords: Mobility transitions; regulation; monopolies; privatisation; competition; co-production

Introduction

Urban transport systems have many features that distinguish them from a classic 'market'. These include the natural monopolies of railway or toll-road networks, and the considerable public investments required for infrastructure and operations. This means that policy approaches derived from neo-liberal theory have limited value. Decisions about urban transport systems are politicised around issues such as environmental impact, equitable access to the benefits of city life, 'best value' for taxpayers, and the fair allocation of public space. These are matters that can be neatly summarised as the 'public good' although this is a slippery concept. Here, it is taken to be the objectively definable benefits that citizens might expect governments to deliver.

The entry of new private interests in communications and information technology, as well as the commercialisation of user-data, necessarily raises new questions about resisting monopolies, directing public investment and maintaining the public good in future urban transport systems. How will public ambitions to improve the social and environmental performance of our transport systems be met as these new and powerful interests deploy emerging technologies?

This chapter uses recent interviews with public and private-sector actors in Australia to explore both obstacles, and the opportunities for these two groups to come together to consider planning and regulatory responses to emerging technologies. In doing this, we want to go beyond simplistic binaries of the 'ossified public sector' and the 'nimble private sector'. We draw from existing literature on collaborative innovation and co-production which allows us to consider how productive engagement between the public and private sectors might help maintain or extend the public good during the current period of technological transition.

In the European context, where to some degree, the public sector remains a strategic leader, manager and regulator of existing urban services and infrastructure, co-production can take place through mechanisms of 'collaborative innovation' (Agger & Sørensen, 2018; Torfing, 2018). An example can be seen in the emergence of Mobility as a Service (MaaS) in Helsinki (Hietanen, 2014, 2019). In Australia, however, where modes of urban governance are increasingly corporatised (Gleeson, 2018) and the state has been retreating from its historic role as a supplier of transport goods and services since the 1990s, the balance of power has become increasingly complex. Watson (2014) provides a useful guide to ensuring that imbalances of power are not ignored in attempts to find new approaches to communicative or collaborative planning; and the literature on co-production points to ways of imagining processes 'through which inputs from individuals who are not 'in' the same organisation are transformed into goods and services' (Ostrom, 1996, p. 1073) while acknowledging inequalities in power.

Our exploration of the options for state responses to manage new 'disruptive' technologies is mindful of unequal power relationships between governments, citizens and corporations and the potential for imbalances in sharing the benefits and risks of innovation.

To probe these issues and to draw out themes for future research, interviews were conducted in 2017 and 2018 with senior staff in public and private sector organisations with interests in the deployment of MaaS and Autonomous Vehicle (AV) technologies in Australian cities.

We had two overarching objectives:

(1) To identify policy instruments that governments or industry might consider appropriate in managing the introduction of new transport technologies. Policy instruments are defined here as specific actions such as policies, contracts, or regulations that are designed to constrain or shape the way new technologies are deployed. Reardon (2020, Chapter 8 in this volume) provides a useful framework to describe how policy instruments fit into the broader process of policy development and implementation.
(2) To use the responses to the interviews to assist in exploring opportunities for the academic community to engage with government and industry practitioners in supporting processes, which lead to the creation of effective policy instruments.

Rather than approach these objectives directly, we organised our interviews around two practical questions that are of immediate concern to participants. First, we asked about the degree to which respondents considered that new technological developments might lead to undesirable monopolies over the platforms to be used to access and pay for new and existing transport services. Second, we asked respondents their opinion on the ways in which the emergence of new technologies might de-stabilise existing transit operations. This is of particular interest in cities like Melbourne, where these operations are governed through complex contracts with international corporations (Department of Transport Victoria, 2019).

The answers to these questions, looked at in the light of our wider objectives, allow us to consider what lessons the Australian experience might have for other jurisdictions about the role of regulation and the nature of public and private sector cooperation. They also allow us to frame clear research questions for future case studies of practice in other cities around the world. Docherty's (2020, chapter in this volume) on multi-level governance provides some useful insights relevant to the complexities of managing technological transitions under Australia's federal system.

We see our research as part of the international response to calls from Marsden and Reardon (2017) and Dowling and Kent (2015) to understand how governance arrangements can and should be shaped by new transport technologies and new modes of interaction between the public and private sectors.

Research Approach

The first series of six interviews was conducted in March and April 2017 with the public-sector staff. These respondents had responsibility for shared and AV mobility policy at managerial level within road, planning and regulatory agencies across the Australian public sector. The second series of interviews was conducted in late 2018 with private-sector staff. Participants were selected from companies with an existing stake in the urban transport sector or in the deployment of new technologies and so liable to some degree of risk from the emerging disruptions. These individuals also had some experience of interactions with state agencies on relevant issues.

Interview subjects were initially identified through the researchers' networks and subsequent 'snowballing'. Our combined sample size of eighteen is considered sufficient for thematic saturation (Guest, Bunce, & Johnson, 2006). However, it was not easy to recruit participants. The political nature of these topics meant that many people we approached were not willing to go on the record: an obstacle that other researchers have also faced (Guerra, 2016). To get agreement to participate, rigid participant anonymity was necessary. This included not only direct identifiers, but also identification by association or location.

Since the aim of this second phase of our research was to compare the opinions of actors from the private sector with those previously obtained from public-sector transport planners (Legacy, Ashmore, Scheurer, Stone, & Curtis, 2018; Stone, 2018), the results were analysed in relation to the key themes from the public-sector interviews rather than through an agnostic process of thematic coding. This places the work within the sphere of confirmatory thematic analysis (Braun & Clarke, 2006).

The two key themes were:

(1) the extent to which regulation is thought to be necessary;
(2) obstacles and opportunities for public-private cooperation to identify and maintain the public good.

Findings

Public Sector Perspectives

Some public-sector respondents clearly understood the need for new technologies to be deployed in ways that strengthen transit systems as an alternative to the private car, and that this would require new responses from governments. Recognising, however, that much of the necessary innovation was being shaped by the market, and there had been a considerable loss of skills over decades from the state to the private sector as a consequence of neo-liberal policies, some doubted the ability of the state to shape developments by planning and public policy methods that have been in place for many decades.

All participants noted the need for state-driven regulations to ensure safety, but some preferred that other regulatory issues be managed through a 'light touch' from government in response to any emerging problems. They recognised that, within the public sector, the skills needed to critically appraise the benefits of the new technologies, and how to procure them as part of an integrated solution, were underdeveloped, because of both skills leakage and the nature of the technologies themselves. It would, in most cases, be unreasonable to expect that many civil servants would understand the technical nuances of the emerging disruptors:

> The capacity of the public sector to understand the drivers that motivate private sector behaviour is going to be important .... A lot of the expertise is going to be held by the private sector, we need to build our capacity to be an informed consumer of services and advice (and) when the private sector holds expertise that the public sector doesn't, it can be very challenging to drive good ... outcomes. [Public Sector interviewee 1]

Interviewees argued that it would be difficult to regulate emergent markets to prevent monopolies emerging. Existing plans were becoming obsolete and new regulations are slow to move through the political process. Many felt that there was considerable risk of regulating 'inappropriately' in an evolving market, where resources were finite:

> Things are moving from (plans that were) written two years' ago – things have changed markedly. The planning cycle and the planning horizons have gone even more crystal ball than they ever were before. It's very hard to plan in such an environment. Regulation takes a long time to design and get in. If you try and foresee what's going to happen and try and regulate ahead of the curve, then you are probably going to get something wrong. [Public Sector interviewee 5]

Although many respondents did feel that governments should leave the field free for private-sector innovation, some did hold a contrary position. One interviewee argued that many in the private sector were looking to government for frameworks in which businesses could operate by setting conditions in which risks could be managed:

> I think that the (companies developing new technologies) are very aware that they need the active cooperation of governments, and (are) probably more (aware) than we as government ... that they need this active cooperation. There are a number of examples – for example, with the trial that Volvo are doing in Gothenburg ... the data, the infrastructure ownership, the rules under which people can use infrastructure ... all of those things require very active and informed governments to be purchasers, regulators, policy makers and partners. [Public Sector interviewee 2]

This pattern was also present during the development of MaaS in Helsinki, where the platform was first created by a state agency before being rolled out as an anti-monopolistic private body, sitting between the operators and the regulator (Hietanen, 2014).

This observation of the recognition of need for 'active cooperation' between government and technology companies was confirmed in the private-sector interviews.

Private Sector Perspectives

As noted above, two key themes emerged from the public-sector interviews: reluctance to regulate digital and technological disruption and consequent obstacles to forming partnerships with industry to manage the deployment of new technologies. Before looking in more detail at how the private-sector respondents viewed these questions, it is worth noting an interesting contrast between the two groups. On one hand, public-sector interviewees recognised the need to regulate for the common good (although recognising the need did not necessarily equate to knowing what should be done) but were largely focussed on limiting intervention in order to allow innovation to flourish. On the other hand, many (although not all) private-sector respondents were looking for a more rule-constrained environment in which to build their businesses.

In this section, we summarise the responses on each of these themes.

The necessity for regulation? The difficulties of regulating in an environment in which new developments are constantly occurring were acknowledged by all respondents. It was not clear how the emphasis, noted by Hensher (2017), on offering the lowest price for services in a typical competitive-tendering market could account for emerging risks to both purchaser and supplier. Introducing new regulations was seen to be especially difficult under long-term franchising arrangements, such as those in Melbourne, where significant parts of transit operations are currently outsourced under contracts which may not be competitively priced and where problems of a lack of transparency and the potential for internal cross-subsidies mean that pricing may not be competitive (Ashmore, Stone, & Kirk, 2018).

The views of respondents on the appropriate response to these difficulties fell along a spectrum. At the one end, some thought that governments should step back to allow innovation and flow.

> […] we've seen time and time again that businesses can thrive and develop more innovative solutions if there is flexibility … and they have productive and cooperative working relationships with regulators … the private sector has the confidence to roll out innovative technological solutions in those markets. (Private Sector interviewee 5)

Other participants felt that the public and private sector needed each other to work towards the common good drawing upon their relative spheres of influence.

Contrary to what one might expect, not every private-sector interviewee put short-term profit maximisation ahead of a need to find sustainable solutions. The undesirable consequences of short-term profit maximisation were noted:

> I tend to, perhaps, not have the greatest confidence in pure private-sector solutions. I think that's just my own experience. Dracula and the blood bank, yeah, I don't think that goes terribly well. So, I guess I'd lean more towards the state-regulated solution. What precisely that should be obviously is being worked out, it's a complex question. (Private Sector interviewee 4)

An ancillary theme was the view that the state was unable to predict every eventuality and should or could only step in when things failed and the public good was compromised. This was seen as the historical pattern for advances in transport technologies:

> [...] the railways were innovative in their day. The private automobile was innovative in its day. So, innovation and disruption are very transitory periods for new technologies and, eventually, the regulatory regime catches up. To some extent, the government doesn't know where it needs to intervene because the normal operating patterns haven't been established. (Private Sector interviewee 8)

As also seen in the public-sector interview, respondents called for 'pragmatism' in response to innovation but could not say what form responses should take, nor who would dictate at what point regulation might be necessary. The notion of the transfer of both commercial and political risk was discussed, with governments seen to 'wear' political risk in event of system failure, regardless of the content of a contract:

> While some public servants might see themselves as policy entrepreneurs and see themselves shaping government policy, that's probably the minority .... The dilemma for government ... is the impact of unintended consequences. Private sector can wear that. If they're successful, they become wealthy. If they're not successful, they fail. Nobody is concerned with a private-sector company failing. It's seen as part of the process. The government, on the other hand, doesn't have that luxury. A failed government program is generally regarded as a bad thing with political consequences, so they become highly risk-averse .... It then puts them in a position of always 'catching up'. (Private Sector interviewee 8)

Interviewees saw several interconnected reasons for governments to step back from imposing regulations. These included the complexity of the issues and the desire of politicians to be seen to be leading the introduction of new technologies ahead of competing jurisdictions. Other commentators have argued that such pressures are compounded by short electoral cycles (three years for Australia's national government, up to four years for state administrations which hold the key responsibilities for urban transport systems), which necessitate quick wins and often discourage long-term strategic planning practice (O'Flynn, Vardon, Yeatman, & Carson, 2011).

While the role of regulation was a dominant theme in the interviews, the need for state planning and guidance through the planning process was also raised. In line with themes flagged in the development-industry press (e.g. Howell, 2018), one participant argued that governments had a role to play in helping guide developers on the transport infrastructure they would need to provide as transport technologies evolved:

> The planning system must have a role in what it allows to happen, and what developers are required to build and provide. At the same time, the planning system is there, to an extent, to protect the surrounding community as well, and to ensure that (things like) car parking are provided appropriately, and managed appropriately … how do we facilitate a shared-mobility outcome now? What is facilitating that shared-mobility outcome in 15 years' time? The trickiest question is, how do we transition between the two? (Private Sector interviewee 6)

Maintaining the public good in an atmosphere of partnership – co-production. Interviewees, both public and private, understood that partnership between the government and the industry was inevitable given private sector control of technological innovations. Some interviewees saw such partnerships operating at the level of data-sharing as a spin-off from the capabilities of new vehicle technologies:

> There has to be some partnering there between government and the holders of data. I know that (cars are now) very sophisticated, they read the road surface, so they know where the potholes are, they know the state of the roads. That is extremely useful to government in terms of maintenance, and if (cars are) collecting that data anyway why would the taxpayer want to pay governments to go around and check the safety of the roads. We can save that cost by pulling in data from the private sector. We can only do that if there's a good relationship between the private sector and the regulator. (Private Sector interviewee 12)

The next level of partnerships discussed by the private sector concerned standardisation of systems. This is challenging in the public policy context. When the data for the common good and sound planning lie outside the government, how does the government ensure access? Might this be used as leverage in regulatory discussion: could the state become actively involved in data ownership as a condition of market access? A degree of 'give and take' was seen to be necessary. As in the often-cited 'Betamax/VHS' battle for the hardware market via data standardisation, it was felt that this would be won by might rather than right: the winner would be the one who saturated the market thereby gaining critical mass long before the state could implement a 'plug and play' standard. This might not be in the wider public or commercial interest:

> If you went back a couple of decades there were those stand-alone operating systems, and gradually there's been (some) standardisation … the issue is: did anyone build up a monopoly unfairly? Well the argument might be no, but now having a monopoly … are you now dominating the market and stopping any competition arising or crushing what's left, and the answer to that might be yes. [Private Sector interviewee 2]

So, there was recognition of potential problems, but no clear direction towards a solution.

Beyond these instrumental partnerships, there is a perception that states are not always able to engage in deeper partnerships to shape the deployment of new technologies to meet wider objectives due to lack of government capacity, orientation, or skills. It was suggested that the 'let the market decide' approach might be a smokescreen for inaction:

> There is a certain sense to (letting the market decide), but I also think that it is a big cop out. I think governments should be saying: we don't know what's going to happen, but we know what we'd like to happen. Instead, it is just too hard … We'll talk about … the fact that technology is too hard to predict and we don't want to get in the way, (but) … that is just code for doing nothing. (Private Sector interviewee 2)

The 'nimble government' desired by most interviewees was seen as unlikely to emerge given current bureaucratic and democratic processes. It was noted that the public service contained deeply competent and knowledgeable people, fully conversant with the issues at play, but their hands were to some degree tied, because if they alerted politicians to the difficulties involved and the contradictions at play, it could ruin their careers:

> I would not give them an A, B or C if I was doing high school scoring with their planning. I feel sorry for some of the people in the state legislature that get it but are surrounded with a system that doesn't want that sort of alternative thinking. It's too hard to deal with … the realities that may be coming in a decade or so – often it's scary and not necessarily a good news story. … if people in government did question the status quo I am pretty sure it would be career limiting. Their political masters would likely go and find someone less insightful for their role. (Private Sector interviewee 7)

This political and operational weakness of Australian transport agencies, described in many interviews, was seen by some as an obstacle to effective regulation.

> I think governments are still floundering ... I don't think that the ride-hailing situation has really taught them anything. I know some State governments thought that they did a good job because they just got out of the way and let it happen and regulated around it, but I think that there are some people in the industry who would take a different view. (Private Sector interviewee 10)

Given this weakness, it will be important to look outside Australia to cities with relatively more powerful city governments and transport agencies for models for engagement with the private sector and the development of appropriate policy instruments.

Useful examples can be found in the responses to the arrival of competing ride-hailing platforms in London (Colley, 2019), Oslo (OECD, 2018) and Los Angeles (Chen, 2019). Each of these cities has tried a different approach to protecting aspects of the public good such as passenger safety and reduced use of public and active transport modes in the face of increasing public acceptance and demand.

Another example is the set of 'ground rules for autonomous driving' being developed by planners at the City of Vienna. These set out the city's traffic management objectives, and state that AVs should be adapted to public spaces rather than the other way around, and that there should be no expectation of public funding for complex infrastructure to accelerate transitions autonomous driving (Stadt Wien, 2019).

As with all policy transfer, however, the recipient nation's capability to successfully adopt the policy is highly dependent upon a variety of localised factors (Stead, de Jong, & Reinholde, 2010; Stone, Stead, Zeibots, Baumann, & Bell, 2014).

Where to Now?

The aim of this chapter was to examine the need, purpose and form of policy instruments that governments or industry might consider appropriate in managing the introduction of new transport technologies.

The interview findings suggest that there is a need for government to regulate emergent technologies in the urban transport sector, and that many in the public and private sectors share similar fears about negative outcomes from emergent monopolies. This is an interesting finding as it has the potential to shift the discussion about regulation away from a binary 'regulate or not' towards a more nuanced, complex and context-dependent discussion about regulation as something that might be co-produced by a range of actors. It also opens the way, in the Australian context of outsourced service and infrastructure provision, to find ways to build flexibility into contracts without additional cost.

Answers to questions of 'what should be regulated' also began to emerge from the interviews.

The rise in platform technologies, and changes to the way people expect to travel, is changing the transport sector in new, exciting, and at times, alarming ways. This research has demonstrated that there are similar concerns surrounding the regulation of this transition. The interviews showed that many participants felt that monopolies are not only undesirable from the perspective of the public good, they are also undesirable for private firms unless they are the monopoliser. There was a sense that both sides were seeking to understand the potential impacts of disruption, even if the values and priorities of different parties were not yet clearly articulated, and often logically contradicted each other.

While it may have been anticipated at the outset that some private sector interviewees may be strongly anti-regulation, many of the interviewees stressed that managing risk in their business entailed maintaining open and competitive markets, and this needed the certainties offered by regulation. 'Disruptive' business would disrupt existing and stable business models, as well as public policy. This is an interesting perspective, as there may in some case be a temptation to see the private sector as monolithic – this preliminary research suggests that this is not the case. Interviewees noted the need for the state to oversee but not necessarily *do*. As an example, in some of the transit franchises operating in Australia, the state retains ownership of all the assets. So, it is possible to replace operators and change contract conditions at regular points of tendering (typically every 7 or 15 years). Franchise change has had various causes, some political, and some the result of inappropriate allocation of risk at the bidding stage (Ashmore et al., 2018; Stone, 2010). This has happened on several occasions, but at no point were the franchisee's revenue or patronage streams dramatically disrupted over the life of their contract, as could be the case under emerging technology transitions, if, for example, there was rapid growth in use competing ride-sharing services.

Perhaps unsurprisingly, questions were raised about the definition of the public good: is it social or economic? This mirrors the political discussions that take place in many societies. *Some* private-sector actors sought to narrow the definition of the 'public good' to the commercial good. This is clearly a concern when the profit imperative is the rationale for decisions about coverage and service levels for ride-sharing in suburban locations. Without regulation or formal contracts in place, the most vulnerable may often remain in conditions of transport disadvantage. This is likely to be the case on the urban fringe, where emergent technologies are unlikely to be able to cater to this need without some form of public subsidy. Models for allocating and monitoring such subsidies are likely to differ substantially from those used for route-buses. This is an important topic for further international and local research and debate. How can ride-hailing firms be contracted to supply definitive off-peak services when they are demand responsive, and in many cases do not claim to employ people? Attempts to establish public-sector alternatives to commercial ride-hailing in suburban markets in Berlin and Vienna have struggled to be competitive.

The basic mechanism of 'market forces' and its strengths and weaknesses were common themes among both public and private sectors. Some respondents relied on the assumption that all aspects of the market mechanism would manifest themselves in the price: as is the case in the classic 'externalities' argument relating to congestion charging (see Powell, 2001). There was discussion of market forces regulating the pricing mechanism, but the evidence from experience is that this has not always proven effective: it is unlikely that competition *in* the market produces sound public outcomes if there are loopholes at play (Ashmore & Mellor, 2010). When some companies can afford to offer 'loss leaders' (products sold at a loss to attract customers), competition can be supressed for a short period until monopolies emerge, and then prices can be raised. Furthermore, in maintaining a competitive market different firms have different lines of capital to draw upon, each with their own distinctive short and long-term imperatives. This can further distort the competitive landscape. In statements about the effectiveness of market forces, a significant proportion of respondents saw monopolies as a potential problem: monopolies are only *not* a concern for the entity likely to form the monopoly.

Again, this was a problem that was stated without a clear direction to a solution. Clearly issues of market power are fundamental to the co-production of workable means of avoiding monopolies.

The similarities in the positions of our interviewees in the public and private sectors add an important dimension to critiques of the Australian state as an increasingly corporatised entity. Stereotypes of a rigid inefficient state and a private sector seeking to escape its restrictive clutches were, in our context, not borne out. This suggests that the planning of future transport provides an opportunity for co-production or co-design of regulation, strategic planning and future policies. Transparent debate will be important to ensure that such co-production is aligned with the values of public-oriented planning systems to produce an integrated system of mobility and access into the future.

This debate needs to start from the reality identified through our interviews. Public and private actors share some understandings of the issues at stake, but because of institutional constraints, the difficulty to track the rapidly changing trajectories of commercial development of new technologies, and limited conceptions of potentially useful policy instruments, neither group appears able to do more than articulate the problems.

International academic collaborations can play a vital role in describing the range of policy instruments being proposed and enacted in different jurisdictions and contextualising their relevance for other cities. Engaged academic research can also help to create processes and structures to bring the public and private sector together as agents facing similar dilemmas and to create the conditions for effective co-production of workable solutions. This is a complex undertaking, with the state expected to be an enabler and an umpire, and the private sector likely to be in favour of competition until they are the ones controlling the market.

Acknowledgements

The authors would like to thank all the anonymous interviewees for the time and thought that they contributed to our conversations. The research was supported by funding from the Melbourne Sustainable Society Institute at the University of Melbourne, and from K2, the Swedish Knowledge Centre for Public Transport.

References

Agger, A., & Sørensen, E. (2018). Managing collaborative innovation in public bureaucracies. *Planning Theory*, *17*(1), 53–73.

Ashmore, D., & Mellor, A. (2010). The 2008 New Zealand public transport management act: Rationale, key provisions, and parallels with the United Kingdom. *Research in Transportation Economics*, *Reforming Public Transport throughout the World*, *29*, 164–182.

Ashmore, D., Stone, J., & Kirk, Y. (2018). The need for greater transparency when assessing the performance and prospects of Melbourne's rail franchise contracts. *Urban Policy and Research*, *37*, 82–96. https://doi.org/10.1080/08111146.2018.1486296

Braun, V., & Clarke, V. (2006). Using thematic analysis in psychology. *Qualitative Research in Psychology*, *3*, 77–101. https://doi.org/10.1191/1478088706qp063oa

Chen, M. (2019). Uber and Lyft hate this bill. Retrieved from https://www.thenation.com/article/uber-lyft-california-ab5/. Accessed on January 24, 2020.

Colley, J. (2019). Uber's troubles in London are nothing compared to the bigger picture. Retrieved from https://theconversation.com/ubers-troubles-in-london-are-nothing-compared-to-the-bigger-picture-127746, Accessed on January 24, 2020.

Department of Transport Victoria. (2019). Melbourne's train and tram contracts. Retrieved from https://transport.vic.gov.au/getting-around/public-transport/train-and-tram-contracts. Accessed on December 9, 2019.

Docherty, I. (2020). Crafting effective policy instruments for 'Smart Mobility': Can multi-level governance deliver?. In A. Paulsson & C. H. Sørensen (Eds.), *Shaping smart mobility futures. Governance and policy instruments in times of sustainability transitions* (pp. 57–73). Brinkley: Emerald.

Dowling, R., & Kent, J. (2015). Practice and public–private partnerships in sustainable transport governance: The case of car sharing in Sydney, Australia. *Transport Policy*, *40*, 58–64.

Gleeson, B. (2018). The metropolitan condition. In S. Hamnett & R. Freestone (Eds.), *Planning metropolitan Australia* (pp. 185–211). Abingdon: Routledge.

Guest, G., Bunce, A., & Johnson, L. (2006). How many interviews are enough? An experiment with data saturation and variability. *Field Methods*, *18*, 59–82.

Guerra, E. (2016). Planning for cars that drive themselves: Metropolitan Planning Organisations, regional transportation plans and autonomous vehicles. *Journal of Planning Education and Research*, *36*(2), 210–224.

Hensher, D. (2017). Future bus transport contracts under a mobility as a service (MaaS) regime in the digital age: Are they likely to change? *Transportation Research Part A*, *98*, 86–96.

Hietanen, S. (2014). Mobility as a Service—The new transport model? *Eurotransport*, *12*, 26–28.

Hietanen, S. (2019). A brief history of MaaS global, the company behind the Whim app. Retrieved from https://whimapp.com/history-of-maas-global/. Accessed on December 2, 2019.

Howell, M. (2018). What property developers need to know about the future of car parking. Retrieved from https://theurbandeveloper.com/articles/what-property-developers-need-to-know-about-the-future-of-car-parking. Accessed on September 21, 2019.

Legacy, C., Ashmore, D., Scheurer, J., Stone, J., & Curtis, C. (2018). Planning the driverless city. *Transport Reviews*, *39*, 84–102. https://doi.org/10.1080/01441647.2018.1466835

Marsden, G., & Reardon, L. (2017). Questions of governance: Rethinking the study of transportation policy. *Transportation Research Part A: Policy and Practice*, *101*, 238–251.

OECD. (2018). Taxi, ride-sourcing and ride-sharing services – Note by Norway. Working Party No. 2 on Competition and Regulation, Directorate for Financial and Enterprise Affairs Competition Committee, OECD.

O'Flynn, J., Vardon, S., Yeatman, A., & Carson, L. (2011). Perspectives on the capacity of the Australian public service and effective policy development and implementation. *Australian Journal of Public Administration, 70*, 309–317.

Ostrom, E. (1996). Crossing the great divide: coproduction, synergy, and development. World Development, 24(6), 1073–1087.

Powell, T. (2001). *The principles of transport economics*. London: PTRC.

Reardon, L. (2020). Smart mobility as a catalyst for policy change towards low carbon mobility? In A. Paulsson & C. H. Sørensen (Eds.), *Shaping smart mobility futures. Governance and policy instruments in times of sustainability transitions* (pp. 139–151). Bingley: Emerald.

Stadt Wien (2019), Basic positions for autonomous driving. Retrieved from https://www.wien.gv.at/stadtentwicklung/strategien/autonomes-fahren.html. Accessed on September 17, 2019 (in German).

Stead, D., de Jong, M., & Reinholde, I. (2010). West-east policy transfer: The case of urban transport policy. In P. Healey & R. Upton (Eds.), *Crossing borders: International exchange and planning practices* (pp. 173–190). London: Routledge.

Stone, J. (2010). Turning over a new franchise: Assessing the health of public transport management in Melbourne. Paper presented at the 33rd Australasian Transport Research Forum, Canberra.

Stone, J. (2018). Planning for disruptive transport technologies: How prepared are Australian transport agencies?. In J. Stone, D. Ashmore, J. Scheurer, C. Legacy, & C. Curtis (Eds.), *Governance of smart mobilities*, London: Emerald Publishing.

Stone, J., Stead, D., Zeibots, M., Baumann, C., & Bell, K. (2014). *Understanding 'best-practice' in transit planning: the importance of tacit knowledge in policy learning*. Paper presented at the Association of European Schools of Planning (AESOP) Conference, 2014, Utrecht.

Torfing, J. (2018). Collaborative innovation in the public sector: The argument, *Public Management Review, 21*, 1–11. doi:10.1080/14719037.2018.1430248

Watson, V. (2014). Co-production and collaboration in planning, *Planning Theory & Practice, 15*(1), 62–76.

Chapter 7

Experimental Governance of Smart Mobility: Some Normative Implications

Annica Kronsell and Dalia Mukhtar-Landgren

ABSTRACT

New forms of 'smart' mobility have emerged with the advance of information technology. From a public sector perspective, these ambitions have been framed both in terms of innovation and sustainability. The development work of these technologies is in part being subsidised by public actors investing in and funding different types of pilots or experiments in order to 'test' these technologies in what is called a real-life environment. This is part of a larger trend of experimental governance in which smart mobility is an important and a possibly growing part. This chapter offers a conceptual analysis of experimental governance by analysing three underlying assumptions in literature and practice (1) the need for extraordinary solutions, (2) the importance of learning by doing and (3) the necessity of collaboration. These three assumptions are analysed in relation to smart mobility experiments in Sweden, and discussed in relation to public values. The concluding discussion elevates a number of normative implications of using experimental governance as a policy instrument for the development of smart mobility.

Keywords: Experimental governance; public values; collaboration; learning by doing; urban experimentation; testbed planning

Introduction

During the last decade, the prefix 'smart' has developed into a strong rhetorical and legitimating device for different growth schemes (Cowley & Caprotti, 2019), ranging from the more overarching notion of smart cities to smart housing, smart living and – the topic of this volume – smart mobility. Smart mobility is an

Shaping Smart Mobility Futures:
Governance and Policy Instruments in times of Sustainability Transitions, 119–135

doi:10.1108/978-1-83982-650-420201007

umbrella concept used to both label and promote innovative solutions related to digitalisation, information and communication technology (ICT) and platform-based solutions within the transport field (cf. Papa & Lauwers, 2015). To date, a number of different actors are forwarding the development of various smart mobility services, including platforms for sharing vehicles, automated vehicles, or different forms of combined mobility services.

Public actors are currently using different tools and policy instruments in the development of smart mobility, ranging from investigating the legislative conditions for automatisation to funding and facilitating collaboration between actors in the transport sector. This chapter describes and discusses one of these governance processes, *experimental governance*. Experimental governance is a broad concept that refers to governing through the funding of, and sometimes participating, in different forms of experiments, pilots and trials in order to facilitate learning. Within the framework of experimental governance, public actors take on different roles (Castán Broto & Bulkeley, 2014; van der Heijden, 2016). Sometimes they lead these processes, and sometimes they are merely partners or enablers for other actors (Kronsell & Mukhtar-Landgren, 2018). In this chapter, we limit the scope to analysing experimental governance as a policy instrument that public actors employ in order to obtain their policy goals. What kind of policy instrument is it then?

In our interpretation, experimental governance, with its temporary, ephemeral character and emphasis on opening up action space for (often private actors) to test solutions, can be seen as one example of what Lascoumes and Le Galés (2007) describe as 'new public policy instruments'. These are often based in 'less interventionist forms of public regulation' (Lascoumes & Le Galés 2007, p. 13), and have their foundation in softer tools such as information, collaboration and consultation, rather than control and regulation. It concerns governing through 'opening up' processes for other actors to aid in the achievement of local political goals (cf. Mukhtar-Landgren, Kronsell, Voytenko Palgan, & von Wirth, 2019). In this chapter, *experimental governance is conceptualised as an instrument employed to promote or accelerate innovation through testing and development of new types of solutions, technologies and services*. It often consists of public actors funding and/or participating in pilots, projects and experiments within – in this particular case – smart mobility. Examples to this end include national innovation agencies funding a MaaS-pilot, regional authorities applying for funding for a pre-study to test autonomous buses or municipalities conducting an experiment on new forms of travel data.

The *first* aim of this chapter is to unpack and explore fundamental assumptions underlying the usage of experimental governance as a policy instrument. We have teased out three assumptions, and these will be analysed and discussed using examples from the field of smart mobility in Sweden. In addition, we forward the argument that public actors have a specific role in experimental governance. As experimentation is inherently a multi-actor process, there is a tendency to see public actors as simply one of many actors. We instead agree with Bryson, Crosby, and Bloomberg's (2014) more general argument that even though '[w]hat is public is seen as going far beyond government' [...] 'government has a special role as a guarantor of public values' (p. 447). What is meant by 'public values' is of course contested and normative, yet an uncontroversial point of departure is that the legitimacy of public actors rests on a democratic and bureaucratic function (Bryson et al., 2014). In this context, we will not delimit or define certain values, instead we will have a broad discussion relating to more overarching public norms. These include both *process values* such as transparency, accountability but also inclusion and participation, and *substantive values* such as sustainability, equality and social justice (cf. Svara, 2006, p. 955; cf. Fainstein, 2000; Lundquist, 1998). This brings us to the *second* aim of this chapter – to explore which challenges experimental governance, as a policy instrument, pose in relation to public values? This second part will be comprised of a more tentative discussion aiming at opening up and exploring new pathways for further research and critical analysis, rather than a more systematic normative analysis.

The chapter is structured as follows: the next section outlines our theoretical understanding of policy instruments and briefly describes the material used in the analysis. The analysis is then divided in two parts: part I outlines the three assumptions underlying experimental governance as a policy instrument in the promotion of smart mobility, and part II discusses the normative implications of these ongoing processes.

A Policy Instrumentation Perspective on Experimental Governance

In this chapter, we understand policy instruments not as a 'tool kit' of discrete and rational tools but follow Lascoumes and Le Gales (2007) understanding of policy instruments as 'both technical and social' devices that 'organizes specific social relations between the state and those it is addressed to [...]' (p. 4). This means that policy instruments are processual in the sense that they, in implementation, continuously structure policy making in different ways, including which resources are used, which actors are deemed relevant, which roles can they take on (or are given) and which further processes are facilitated or blocked through their application (cf. Lascoumes & Le Gales, 2007, p. 9). In addition, Lascoumes and Le Gales (2007) argue that 'every instrument constitutes a condensed form of knowledge about social control and ways of exercising it' (p. 3). In our interpretation, this means that policy instruments build on certain assumptions, including an expectation on what type of effects will be the outcome of different policy instruments, or as is it formulated by the authors themselves 'understanding *the reasons* that drive towards retaining one instrument rather than another, but also *envisaging the effects* produced by these choices' (Lascoumes and Le Gales, 2007, p. 4, our emphasis). In this chapter, we explore these underlying assumptions in regards to both reasons and envisaged effects. As will be motivated and developed below, we have, from the growing literature, discerned three underlying assumptions that characterise experimental governance as a policy instrument. These are (1) *the need for extraordinary solutions;* (2) *the importance of learning by doing and* (3) *the necessity of collaboration.*

The chapter is structured the following way: we will discuss these three assumptions one at a time, firstly describing how they are currently forwarded or described in the literature on experimental governance, and secondly exemplifying, describing and discussing the assumptions in relation to practices of experimental governance from the field of smart mobility in Sweden. Each section concludes with the elevation of some critical reflections in need for further analysis. In the concluding section, these critical themes will be recapitulated, elevated and discussed further. The empirical material of the analysis from Sweden consists of web pages and policy documents from public actors engaged in funding or participating in different forms of experimentation. These include the Swedish innovation agency Vinnova, and the innovation programmes Drive Sweden and Viable Cities that, in similarity to Vinnova funds a number of projects. We have also included material from a number of different pilots and projects including municipal actors as participants. Here, we have especially focused on the municipality of Göteborg as it is an important node for development and innovation ventures within the Swedish transport sector.

Part I: Experimental Governance as a Policy Instrument – Promoting Smart Mobility in Sweden

In this section, three underlying assumptions that characterise experimental governance as a policy instrument will be described, after the first one on (1) *the need for extraordinary solutions,* we will move to (2) *the importance of learning by doing and* finally (3) *the necessity of collaboration.* These will be developed using academic literature and illustrated with examples from the Swedish context.

The Need for Extraordinary Solutions

In the academic literature, experimentation is often forwarded as necessary in light of both the urgency and the difficulties in addressing current (often climate) challenges (Evans, Andrew, & Raven, 2016; von Wirth, Fuenfschilling, Frantzeskaki, & Coenen, 2019; Voytenko, McCormick, Evans, & Schliwa, 2016). Experiments are, in this regard, often embedded in narratives of 'wicked' problems. The term was coined by Rittel and Weber (1974) and refers to problems that are 'complex, unpredictable, open ended, or intractable' (as cited in Head & Alford, 2015, p. 712). The concept has been used within policy and governance literatures (Head & Alford, 2015) and extensively in the sustainability literature, where climate problems have even been considered 'super-wicked' (Lazarus, 2008). This is due to their persistent nature, their complexity and systemic character (see Frame, 2008; Frantzeskaki, Loorbach, & Meadowcroft, 2012; Wolfram, Frantzeskaki, & Maschmeyer, 2016) and as a result is difficult to solve, presumed even transformational if solved (Wiek & Lang, 2006) and requiring extraordinary means. This in turn suggests that the solutions promoted in experimental governance should be not only novel but also disruptive.

Images and descriptions of smart mobility as disruptive is highly prevalent in our empirical material. Yet, the novelty and disruptive potential of smart mobility is often emphasised without reference to 'wicked problems' in policy documents.

In contrast, the novelty and extraordinary character of the solution is instead depicted as a value in itself.

> It's not all about driverless vehicles. This is a completely new approach to mobility. We are on the threshold of a radical shift, and it's happening fast. In just a few years the world will change. We will see entirely new mobility business models enabling sustainable cities. (Drive Sweden, 2019)

Drive Sweden, quoted above, is one of government's 17 strategic innovation programmes. On their web page, they describe the development of smart mobility as a 'completely new', and 'radical shift' where the 'world will change'. Sustainability is mentioned as the context for this development (although limited to 'sustainable cities'). On a similar note, but using climate change as the point of departure, the innovation agency *Vinnova* state on their homepage that:

> To cope with climate challenges, society needs to use transport resources smarter and more efficiently. We need new solutions that are connected, autonomous and multi-modal within road, sea, rail and air. Sweden has a fantastic opportunity to become the world leader in this area, but we need more players to make contributions. (Vinnova, 2019)

Here, the relation between problems and the 'need' for 'new solutions' is again highlighted; the climate challenges require us to go beyond business as usual. Yet the final sentence is, in our interpretation, perhaps more representative for the policy context of experimental governance, the ambition to become a 'world leader in this area'. This is a second rationality that underpins the need to promote 'extraordinary solutions' – the will to brand the city and/or to promote innovation and growth. Another example to this end comes from the municipality of Lund. At a breakfast meeting hosted by the municipality in 2018 'representatives from the automotive industry, innovation clusters, vehicle clusters, politicians, various traffic operators, researchers and the municipality' participated to discuss the topic 'Steering the driverless society':

> Here in Lund we are going to find new areas to invest in. Smart mobility is one such area, clean tech is another, says the municipal head of business administration Per Persson.
>
> […] - We have to have a test bed for self-driving vehicles in Lund, says Mia Rolf, CEO of Ideon Science Park AB. If we do something unique here we can attract international companies to the city. (Lund Municipality, 2019a)

In the quote, innovation, attraction and growth are presented as the context or motivation behind investing in testbeds and experimentations. This tendency has also been noted in the literature on experimental governance, where research from the municipal level has shown that participating cities embed their ambitions in a broader municipal development work that can be driven by several different motives, including increasing urban attractiveness in relation to tourists and investors while participating in the race to become one of the future 'smart cities' (cf. Berglund-Snodgrass & Mukhtar-Landgren, 2020; von Wirth et al., 2019). Thus, solving sustainability problems is on the agenda in relation to smart mobility, but it is only one among several objectives.

Summing up, these shifting – and often implicit – priorities and motivations open up questions of which policy goals that public actors are striving for in urban experimentation. In aiming for both innovation and sustainability, there is an assumption that these goals are compatible at all times. This is not necessarily the case and opens up for the responsibility of public actors to specify which goals they strive for, and cater experiments as to obtain those goals. In addition, it opens up questions regarding 'smart cities' as a public goal. What are the merits of striving for smartness for public actors?

Learning by 'Doing'

This section discusses the second assumption: the importance of learning by doing. At its core, experimentation is 'a process that generates learning through an explicit intention to test new ideas' (McFadgen & Huitema, 2016, p. 1765). Often it is the city that provides an arena for learning where different types of public and private actors meet (Laakso, Berg, & Annala, 2017; Sengers, Wieczorek, & Raven, 2019) and where different kinds of experiments are conducted. One example is the public funding of pilots and projects, and another example is the urban testbed, which has been defined as 'a geographically delimited site of urban development, in which urban experiments constitute an integral part of planning and developing the area' (Berglund-Snodgrass & Mukhtar-Landgren, 2020). Yet another example is *Urban Living Labs* (ULLs) defined as 'sites in cities devised to design, test and learn from social and technical innovation in real world settings' (Voytenko et al., 2016), or as a specific type of learning device in cities 'which are intended to design, demonstrate and learn about the effects of urban interventions in real time' (Bulkeley et al., 2016, p. 13). As the quotes indicate, notions of 'real time' but also 'real world' are elevated in the definitions, hence the notion of *living labs*. Labs are often seen as examples of new developing forms of urban governance (Menny, Voytenko, & McCormick, 2018; Smas, Schmitt, Perjo, & Tunström 2016). These processes are named laboratories because of the association with testing and learning. In the literature, laboratories have also been distinguished further according to the degree of control over the experiment. Bulkeley et al. (2019) discuss that it can function as a trial, which is controlled regarding the particular outcomes, as an enclave when the boundaries of the experiment are assured, as a demonstration, which provides a showcase of what the space may look like and finally as a platform, which is about creating conditions for multiple new relations and arrangements. In addition to testing in real life, there is one more aspect of this assumption, and that is the notion that a successful experiment (successful in terms of speaking and responding to the problem) should be able to scale up, spread and be reproduced in new contexts and possibly also in response to new problems (see e.g. Sengers et al., 2016; Williams, 2016). In other words, the ambition is to share experiences to facilitate broader processes of policy learning and knowledge dissemination (Evans et al., 2016).

The notion of 'learning from doing' is highly prevalent in the material, and in many ways provides a core legitimising logic behind the funding and enabling of pilots and tests for smart mobility. In addition, it takes on different forms. In relation to the example of public funding of pilots and projects, *Drive Sweden* funds a number of more delimited pilots annually, ranging from tests of business models to technological solutions. In addition, we have a number of testbed planning processes in Sweden. One example is Barkaby in Stockholm, a development site where an autonomous minibus is being tested, and another is Brunnshög in Lund where a MaaS-solution integrated in a new, parking free, apartment building is being built and piloted. In Sweden, we also have a prevalence of platforms for testing. One such example is *ElectriCity.*

> As part of ElectriCity we are also creating a platform for the development and testing of services and products that can contribute to more attractive public transport [...] The idea is that these should be able to be scaled up outside the demo arena (Electricity, 2019a).

As indicated from the quote, several components are visible here, from platforms to testing and scaling up potential solutions. Another example is 'Realitylab Göteborg'. Under the description of the lab – with the subheading 'analyze – discover – develop – deliver: in a real environment with real citizens of Gothenburg', the municipality describes how:

> The new technologies around self-driving, shared, electric vehicles in system solutions need to be tested in the complex reality, in order for us to understand how they can best be implemented. By 'reality lab' we mean that tests are done in 'ordinary environment' with 'ordinary users' so that technology, business models and regulations are developed, tested and demonstrated in a 'real' context (our translation). (Göteborg Municipality, 2019a, quotation marks in original)

One important aspect of the discourses on experimental governance is the necessity of 'real life' environments to test solutions. Testing in real life entails placing the test in a situation that mimics the context, in which the test will be scaled up and spread. This is also verbalised in the testing of autonomous vehicles in Gothenburg, using concepts such as 'everyday context' and 'real life':

> It is important that the technology is optimized for real traffic conditions. Starting in December 2017 we will conduct a unique pilot where families in the Gothenburg region will be able to test an autonomous vehicle, in an everyday context in real life (our translation). (Göteborg Municipality, 2019b)

Here, the quote refers to machine learning and that the algorithms need real life situations in order to gather information. Algorithms and machine learning are important aspects of smart mobility, but it is not the only reason for emphasising real life testing. *What* is it that is tested? And what *type* of knowledge is sought after? The following quote exemplifies what type of solutions is tested:

> For example new bus stop solutions, traffic management systems and safety concepts as well as systems for energy supply and energy storage. What is more, new business models for sustainable mobility in the city will be tested. (Electricity, 2019b)

The quote indicates that what is being tested is primarily technological solutions. This has also been indicated in previous research on smart mobility experiments, where there is a tendency to emphasise the specific technology being tested, rather than the broader societal goals (Berglund, Mukhtar-Landgren, & Paulsson, 2019; Manders, Wieczorek, & Verbong, 2018). Even though they are often implicit in the goal of making public transport more attractive, the association with goals like decreased traffic congestion, emissions or increased safety and more liveable cities is generally downplayed in relation to innovation and competitiveness.

Summing up, we have shown that experimental governance as a policy instrument is strongly based in the assumption that learning is expected to be enhanced when it takes place in a kind of laboratory that is set in real-life. To what extent can the municipality control this development and assure that it benefits society or the municipality more generally and not only as support for a specific technology? Another question that arises is what kind of real life is it that is envisioned in the smart mobility testing? Whose everyday context is it deemed interesting to experiment with?

The Necessity of Collaboration

Research on urban experimentation has also highlighted the centrality of collaboration between different needs owners (Menny et al., 2018; Voytenko et al., 2016). Previous empirical studies have shown that collaboration can be carried out and organised in different ways, ranging from informal, loose collaboration such as a workshop to formally organised collaboration with contracts and steering groups (Mukhtar-Landgren et al., 2019). There is generally an assumption that the complexity (mentioned above) requires the knowledge and expertise of several actors. Even though this is true for several policy areas within which experimentation is taking place, certain smart mobility solutions, such as combining different services in one solution (such as MaaS), are intrinsically collaborative (Mukhtar-Landgren & Smith, 2019), and thus collaboration is a necessary prerequisite for the service to be launched. Collaboration within the frame of experimentation is associated with different related concepts, which indicate a variation of collaborative design and formality. In the literature, we find concepts such as co-creation, co-production, quadruple helix-model, and multi-stakeholder involvement (see e.g. Franz, 2015; Leminen, 2013; Schuurman & De Marez, 2012) with set-ups including actors from the public, private, academy as well as users, and where the importance of including citizens and users is often highlighted (Marvin, Bulkeley, Mai, Mccormick, & Palgan, 2018).

Collaboration is highly relevant to experimental governance in the Swedish context of smart mobility. It is illustrated continuously and consistently throughout our material. On Drive Sweden's webpage, it is noted that:

> [...] in order for this to happen we need to work in a cross-functional way. For example, a vehicle manufacturer cannot trigger this development only by itself; neither can a city facilitate the transformation of mobility without working closely with lots of other stakeholders. (Drive Sweden, 2019)

In essence, experiments in relation to mobility consist of partnerships between different actors: the players who test a solution in the project or pilot, and different stakeholders who contribute to the experiment with their experience, insight or in the capacity of potential future users. Although collaboration can be bottom-up initiatives emerging from the actors and their networks, collaborations are also initiated by funding institutions. For instance, many of the organisations at hand, such as Viable Cities and Drive Sweden, are in themselves collaborations, and in addition, these organisations also start new organisations that are – again – collaborative. For instance, Drive Sweden is a collaborative organisation, which in turn hosts the collaborative organisation KOMPIS. Below we find three different collaborative groups that also collaborate with each other:

> KOMPIS is initiated by the Government's collaboration group for Next Generation Travel and Transport, and is a project under Drive Sweden, funded by Vinnova (our translation). (Kompis, 2019)

This way of starting up collaborative organisations, or 'getting people in the same room' is, as mentioned above, a noted policy instrument both within experimental governance and within the field of smart mobility (e.g. Karlsson et al., 2020; Mukhtar-Landgren & Smith 2019). Besides these state initiatives, municipalities also arrange collaborative ventures such as workshops and conferences to promote experimentation. One example was the breakfast meeting from Lund mentioned above, and another example was the event *Tomelilla innovation week* on the theme 'Digitalization, innovation and future travel' in May 2019. In addition, municipalities are important, even pivotal, collaborative actors in experimental governance in smart mobility in different ways. They participate as key actors in a range of different collaborations, thus they do not only initiate collaborations, but they also participate in them as partners. Finally, another way that public actors promote collaboration (that they perceive of as necessary) is through explicit requirements for collaboration in their call for projects.

Summing up, there is a strong emphasis on, and assumption of, the necessity to collaborate in experimentation on smart mobility. In terms of normative implications, collaborations raise issues related to both transparency and accountability in decision making, as well as inclusion of citizens. One aspect that has been particularly highlighted is the tendency to primarily include citizens in the capacity of users.

Part II: Experimental Governance as a Policy Instrument – Normative Implications and the Role of Public Values

In this final section, we will elevate the normative implications hinted to previously and discuss them in relation to previous research on the development of smart mobility and in relation to broader debates on experimentation and innovation.

The Need for Exceptional Solutions – Some Normative Implications

Experiments, in the ideal typical understanding of the word in the context of experimental governance, are about finding solutions to difficult, or even so called 'wicked' problems. In relation to sustainable passenger transport, these challenges are about car dependency, which in turn concerns challenges such as traffic congestion, the proliferation of parking garages that affect the city's green spaces, but also noise and dangerous emissions from exhaust fumes. These relate to important public values and are today often high on the agenda of many municipalities, but they are also highly politicised issues as they ignite public protest from groups anxious to hold on to their car and their parking space.

Above we showed how smart mobility was claimed to be – not a solution to an *extraordinary problem* – but rather an *extraordinary solution in its own right*. This raises the question if 'smart' can be considered a public goal that is to be pursued by public actors? The question of whether or not public actors should pursuit smart solutions is a normative analysis in its own right, beyond the scope of this chapter. Yet some points can be raised: Smart mobility builds on previous developments in transport planning, with a shift from a car-based planning – with speed and mobility as core values – to people-based planning expressed as sustainable mobility (Banister, 2011; Hall, Gudmundsson, Marsden, & Zietsman, 2014) with core values being land-use and accessibility. In this sense, it seems to affirm its connection to some public values in the case that these sustainability goals are placed centre stage. However, two features stand out in relation to experimentation and smart city development: firstly, that smart is centred on (information) technology, and the integrity of persons on which data are being collected (Oldbury & Mukhtar-Landgren, 2020), and secondly, on the emphasis on consumers as end-users of the service provided (Papa & Lauwers, 2015). We will return below to the tendency to reduce citizens to 'users' or 'customers', but an important point to make here, in relation to smart mobility, is that of digital information.

As mentioned above in the example from Göteborg, one part of testing relates to machine learning and developing algorithms. Another aspect is that several smart solutions, such as MaaS, but also shared mobility solutions, are dependent on a platform. This type of digital infrastructure is often based in publicly owned data, which raises issues relating to ownership, usage and user integrity. In addition, research indicates that algorithms are not 'neutral' from the perspective of gender and skin colour (Mattern, 2014). Public actors interacting with smart mobility need to take into account that platforms, machine learning and the development of algorithms raise questions in relation to public values, concerning both process values such as transparency, and substance values such as equality in the cases that algorithms leads to an uneven distribution off services (Oldbury & Mukhtar-Landgren, 2020).

Moving on to the results of the empirical analysis, we noted (as mentioned above) that smart mobility was claimed to be an extraordinary solution in its own right rather than a solution to an extraordinary problem. Values relating to climate concerns and the pursuit of sustainability are prevalent and often mentioned in the empirical material, even though they are rarely placed clearly central stage in the descriptions of pilots and experiments. However, the goals elevated in the material were frequently goals related to innovation capacity, growth and competitiveness and (for municipalities) the possibility to create a positive image for the city.

Whether these values can work together and become win-win was studied by Noy and Givoni (2018) who examined the extent to which 'smart' and 'sustainable' are aligned. They asked commercial actors and found that transport entrepreneurs mainly had commercial interests, but these actors were also of the conviction that technological developments such as autonomous vehicles can by themselves lead to sustainable transport. Yet Lyons (2018) cautions us that corporations are exerting their power in smart mobility and in pursuit of goals that may or may not be aligned with the goals of urban planners who have a broader ambition of social and environmental sustainability. There is a potential tension between experimental governance on smart sustainability – particularly regarding its connection to commercial interests and competition – and public values. The relation between goals of smart mobility experiments and the role of public values such as sustainability – but also broader goals, such as social inclusion and equality, merit further inquiries. This includes taking into account that the inclusion of commercial interest in experimentation places demands on public actors to govern towards public values.

Learning by doing – Some Normative Implications

When experimental governance comes in the form of laboratories or test-beds, it appears as if it is limited, defined and controlled. This is elusive. The urban context and the site for experimentation are not in any way an 'empty or isolated space'. This means that ongoing trends in public administration, such as austerity and cut-backs in the welfare sector, conceptual changes on the role of governments, the development towards increasing digitalisation, and a strong confidence in competition and market solutions to handle a perceived lack of efficiency (Bryson et al., 2014, p. 446) will affect the 'landscape' where smart mobility experiments are conducted. These factors will also determine the answer to the question of *what*, more specifically, is being tested? When experimenting with, for example, a MaaS-solution, a number of aspects may be tested, ranging from the business model to the possibilities of forming inclusive and accessible solutions. As mentioned above, research from the field of smart mobility has shown that it is often a specific technological solution that is tested, where context is often delimited to issues such as business models (Berglund & Mukhtar-Landgren, 2019; Manders et al., 2018). Yet we would argue that the early stages of policy development – which experimentation actually is – are in itself an opportunity for public actors to raise public values such as inclusion early on. How can the municipality interfere in smart mobility experimentation in a way that benefits citizens, society and the municipality more generally rather than just supporting innovation initiatives or a specific smart mobility technology? This is a question that merits further research.

Another trait that we explored under this subheading is the notion of learning by *testing in real life situations*. The analysis showed that this was assumed to be conducive to learning in smart mobility, in a way that would lead to the spreading and up-scaling of the experiment. The assumption that experimental governance, in order to be successful, need to conduct tests in real life raises a number of questions. One aspect of real life is related to machine learning to refine algorithms for autonomous vehicles, and other times it is related to testing a MaaS-solution in relation to a workplace. Here, the question arises of who is included or excluded in this real life, whose everyday experience, including working life travel patterns, is reckoned as interesting for learning and experimentation. Here, we know from numerous accounts from gender studies and transport that 'the everyday experience' of men and women's travel patterns differ widely (Kronsell, Smidfelt Rosqvist, & Winslott Hiselius, 2016). These aspects relate again to public values such as inclusion but also to gender equality.

Collaboration – Some Normative Implications

The third assumption of experimental governance is that *collaboration* is necessary. Municipalities increasingly take on the role as enabler of collaboration through experiments that bring together actors that can share risks, contribute to the success of experiments and lead it forward to become scaled-up or commercialised. The municipality is a key actor because of its legitimate position in the local context, its trustworthiness and its role in highlighting priorities for the city through urban visions and plans. It raises questions relating to process values about how transparent and open the process is, something which can become delicate when there are business interests involved, as they are competing on a market, where secrecy rather than transparency is the norm. This can be an added challenge to the fact that democratic values related to transparency and accountability are a key challenge to any collaborative process (e.g. Blomgren-Bingham & O'Leary, 2008; Huxham, 2003). These issues need to be safe-guarded for by the public actors that participate in these processes, and the extent to which this is taken into account by public actors needs further study.

As mentioned above, the smart movement has a tendency to place emphasis on consumers as end-users of the service provided (Papa & Lauwers, 2015). This means that interest in citizens is mainly as users, with a focus on how they perceive the service, how easy it is to use and how it can be developed to be more user-friendly. While the perspective of the citizen is central, this is not the kind of participation and inclusion envisioned for process values and local democracy. We can draw a parallel to urban living labs, where experiments mostly are technical solutions in energy efficiency and transport. Here too, the ambition is to include citizens but when studied, most cases show it has been mainly about engaging with users of the technology rather than allowing for a broader societal critical engagement and involvement. One example comes from Laakso et al. (2017) who studied 90 urban experiments in Amsterdam and found that participation by 'users' was often merely symbolic, only 12 experiments were fully participatory (see also Steen & van Bueren, 2017). If experiments cannot live up the process values of participation and inclusion, it is a concern, and particularly if municipalities are engaging with and investing resources in it.

Concluding Remarks

The chapter discussed experimental governance in terms of a policy instrument based in three key assumptions: (1) the need for extraordinary solutions, (2) the importance of learning by doing and (3) the necessity of collaboration. Focussing on smart mobility we used the empirical case of Sweden to tease out these elements of the experimental governance instrument as they materialise in practice. In accordance with Lascoumes and Le Gales conceptualisation of policy instruments (2007), policy instruments continuously take new forms in implementation. Policy instrumentation processes within experimentation lead to new relations between actors, including new relations between companies and public actors, between public actors and citizens and between different public actors (cf. Mukhtar-Landgren et al., 2019). It also introduces new routines, and when municipalities increasingly engage in experimental governance, for example, by enabling rather than legislating, the authority of municipalities shifts (cf. Kronsell & Mukhtar-Landgren, 2018). While the municipality's capacity to initiate and occasionally govern collaborations through enabling collaboration might potentially deliver mobility solutions that are 'smart', efficient and have low carbon emissions, we ask how they can live up to goals relating to other public values – substantive as well as processual. Taking the processual perspective of Lascoumes and Le Galés into account, a number of critical points were raised through the analysis. These are elaborated below.

The analysis pointed towards some normative and critical implications of smart mobility that merit further analysis. In relation to the assumption of *extraordinary solutions*, it is important to highlight that smart mobility is not an end in itself, but instead a means to solve extraordinary problems, such as the climate crises. In practice, it often becomes just that. Second, we briefly discussed how 'smart' was a policy to strife for, and pointed towards the possible challenges related to ICT and platform technology and the use of data in relation to personal integrity.

Finally, we discussed the notion of multiple goals in the development of these extraordinary solutions, and questioned whether sustainability and competitiveness are always compatible. In regard to the assumption of *learning by doing,* we highlighted the issue of what it is, more specifically, that is being tested, as well as the notion that a 'real life environment' is not a neutral space, but a space that is embedded in power relations. In addition, there is no such thing as a 'neutral gaze' from which experiments can be conducted, not least taking into account the differences in, for instance how men and women move in public space. Finally, the third aspect relates to *collaboration*. Here, we raised two points that are well known in the broader collaboration literature, on the one hand the challenge of transparency and accountability in collaboration, and secondly the issue of inclusion and pointed out the tendency to, in practice, reduce citizens to users.

We end with James Evans (2011, p. 237) who concludes about the experimental city that:

> the central role afforded to experimentation in current manifestations of urban sustainability undoubtedly offers a potential space for more playful or insurgent political engagements with urban infrastructure and material form.

However, he continues that 'if sustainability comes down to letting 1000 experimental flowers bloom, then it matters who gets to experiment, and how'. If governance is being transformed with quatro helix networks of public actors-business practitioners-science-grassroot activists that render traditional distinctions obsolete then we may ask if this also changes the balance of democratic authority and how it may affect how public values are supported.

References

Banister, D. (2011) Cities, mobility and climate change. *Journal of Transport Geography*, *19*(6), 1538–1546.

Berglund-Snodgrass, L., & Mukhtar-Landgren, D. (2020). Conceptualizing testbed planning: Urban planning in the intersection between experimental and public sector logics. *Urban Planning*, *5*(1), 2183. doi:10.17645/up.v5i1.2528

Berglund-Snodgrass, L., Mukhtar-Landgren, D., & Paulsson, A. (2019). Experiment för hållbar mobilitet. Vad innoveras det (inte) kring i svenska kommuner? [Experiments for sustainable mobility: What are Swedish municipalities (not) innovating about?]. In J. Algehed, E. Eneqvist, C. Jensen, & J. Lööf (Eds.), *Stadsutveckling: En forskningsantologi om organiseringsutmaningar för stad och kommun [Urban development: A research anthology about organisational challenges]* (pp. 89–112). Gothenburg: Mistra Urban Futures.

Blomgren-Bingham, L., & O'Leary, R. (2008). *Big ideas in collaborative public management*. New York, NY: Routledge.

Bryson, J., Crosby, B., & Bloomberg, L. (2014). Public value governance: Moving beyond traditional public administration and the new public management. *Public Administration Review*, *74*(4), 445–456.

Bulkeley, H., Lars, C., Frantzeskaki, N., Hartmann, C., Kronsell, A., Mai, L., … Voytenko Palgan, Y. (2016). Urban living labs: governing urban sustainability transitions. *Current Opinion in Environmental Sustainability*, *22*, 13–17.

Bulkeley, H., Marvin, S., Voytenko Palgan, Y., McCormick, K., Breitfuss-Loidl, M., Mai, L., … Frantzeskaki, N. (2019). Urban living laboratories: Conducting the experimental city? *European Urban and Regional Studies*, *26*(4), 317–335.

Castán Broto, V., & Bulkeley, H. (2014). Maintaining experiments and the material agency of the urban. In S. Graham & C. McFarlane (Eds.), *Infrastructural lives: Urban infrastructure in context* (pp. 199–218). London and New York, NY: Routledge.

Cowley, R., & Caprotti, F. (2019). Smart city as anti-planning in the UK. *Environment and Planning D: Society and Space*, *37*(3), 428–448.

Drive Sweden. (2019). About Drive Sweden. Retrieved from https://www.drivesweden.net/en/about-drive-sweden. Accessed on December 11, 2019.

Electricity. (2019a). ElectriCity – Cooperation on tomorrow's transports. Retrieved from https://www.electricitygoteborg.se/en/node/19505. Accessed on December 14, 2019.

Electricity. (2019b). ElectriCity – Cooperation on tomorrow's transports. Retrieved from https://www.electricitygoteborg.se/en/node/19505. Accessed on December 13, 2019.

Evans, J., Andrew K., & Raven, R. (2016). The experimental city: New modes and prospects of urban transformation. In J. Evans, K. Andrew, & R. Raven (Eds.), *The experimental city* (pp. 1–12). London: Routledge.

Evans, J. P. (2011). Resilience, ecology and adaptation in the experimental city. *Transactions of the Institute of British Geographers*, *36*(2), 223–237.

Frame, B. (2008). 'Wicked', 'messy', and 'clumsy': Long-term frameworks for sustainability. *Environment and Planning C: Government and Policy*, *26*(6), 1113–1128.

Frantzeskaki, N., Loorbach, D., & Meadowcroft, J. (2012). Governing societal transitions to sustainability. *International Journal of Sustainable Development*, *15*(1–2), 19–36.

Franz, Y. (2015). Designing social living labs in urban research. *Info*, *17*(4), 53–66.

Göteborg Municipality. (2019a). Verklighetslabb Göteborg [Reality lab Göteborg]. Retrieved from https://goteborg.se/wps/portal/enhetssida/Innovation-och-utveckling-far-framtidens-mobilitet-i-Gateborg/verklighetslabb-goteborg/. Accessed on December 14, 2019.

Göteborg Municipality. (2019b). Drive me. Retrieved from https://goteborg.se/wps/portal/enhetssida/Innovation-och-utveckling-far-framtidens-mobilitet-i-Gateborg/driveme. Accessed on December 14, 2019.

Hall, R., Gudmundsson, H., Marsden, G., & Zietsman, J. (2014). Sustainable transportation. In M. Garrett (Ed.), *Encyclopedia of transportation: Social science and policy* (Vol. 1, pp. 1300–1304).. Thousand Oaks, CA: SAGE Publications, Inc. doi: 10.4135/9781483346526.n457.

Head, B. W., & Alford, J. (2015). Wicked problems: Implications for public policy and management. *Administration & Society*, *47*(6), 711–739.

Huxham, C. (2003). Theorizing collaboration practice. *Public Management Review*, *5*(3), 401–423.

Karlsson, M., Mukhtar-Landgren, D., Smith, G., Koglin, T., Kronsell, A., Lund, E., ... Sochor, S. (2020). Development and implementation of Mobility-as-a-Service: A qualitative study of barriers and enabling factors. *Transportation Research, Part A: Policy and Practice*, *131*, 283–295. https://doi.org/10.1016/j.tra.2019.09.028.

Kompis. (2019). Det här är kompis [This is kompis]. Retrieved from https://kompis.me/#det-har-ar-kompis. Accessed on December 11, 2019.

Kronsell, A., & Mukhtar-Landgren, D. (2018). Experimental governance: The role of municipalities in urban living labs. *European Planning Studies*, *26*(5), 988–1007.

Kronsell, A., Smidfelt Rosqvist, L., & Winslott Hiselius, L. (2016). Achieving climate objectives in transport policy by including women and challenging gender norms – The Swedish case. *International Journal of Sustainable Transportation*, *10*(8), 703–711.

Laakso, S., Berg, A., & Annala, M. (2017). Dynamics of experimental governance: A meta-study of functions and uses of climate governance experiments. *Journal of Cleaner Production*, *169*, 8–16.

Lascoumes, P., & Le Gales, P. (2007). Introduction: Understanding public policy through its instruments. *Governance*, *20*(1), 1–21.

Lazarus, R. J. (2008). Super wicked problems and climate change: Restraining the present to liberate the future. *Cornell Law Review*, *94*, 1153.

Leminen, S. (2013). Coordination and participation in living lab networks. *Technology Innovation Management Review*, *3*(11), 5–14.

Lund Municipality. (2019a). Lund styr mot smartare mobilitet [Lund is governed towards smarter mobility]. Retrieved from https://www.lund.se/foretagare/nyheter-och-nyhetsarkiv/2018/lund-styr-mot-smartare-mobilitet/. Accessed on December 06, 2019.

Lyons, G. (2018). Getting smart about urban mobility-aligning the paradigms of smart and sustainable. *Transportation Research Part A: Policy and Practice*, *115*, 4–14.

Manders, T. N., Wieczorek, A. J., & Verbong, G. P. J. (2018). Understanding smart mobility experiments in the Dutch automobility system: Who is involved and what do they promise?. *Futures*, *96*, 90–103.

Marvin, S., Bulkeley, H., Mai, L., Mccormick, K., & Voytenko Palgan, Y. (Eds.). (2018). *Urban living labs – Experimenting with city futures*. London: Routledge.

Mattern, S. (2014). Interfacing urban intelligence. *Places Journal Online*. Retrieved from https://placesjournal.org/article/interfacing-urban-intelligence/. Accessed on October 23, 2019.

McFadgen, B., & Huitema, D. (2018). Experimentation at the interface of science and policy: a multi-case analysis of how policy experiments influence political decision-makers. *Policy sciences*, *51*(2), 161–187. https://doi.org/10.1007/s11077-017-9276-2

Menny, M., Voytenko P. Y., & McCormick, K. (2018). Urban living labs and the role of users in co-creation. *GAIA-Ecological Perspectives for Science and Society*, *27*(1), 68–77.

Mukhtar-Landgren, D., Kronsell, A., Voytenko Palgan, Y., & von Wirth, T. (2019). Municipalities as Enablers in Urban Experimentation. *Environmental Policy and Planning*, *21*, 718–733. https://doi.org/10.1080/1523908X.2019.1672525

Mukhtar-Landgren, D., & Smith, G. (2019). Perceived action spaces for public actors in the development of mobility as a service. *European Transport Research Review*, *11*, 32. https://doi.org/10.1186/s12544-019-0363-7

Noy, K., & Givoni, M. (2018). Is 'smart mobility' sustainable? Examining the views and beliefs of transport's technological entrepreneurs. *Sustainability*, *10*(2), 422.

Oldbury, K. & Mukhtar-Landgren, D. (2020). Delad smart mobilitet? Tillgång, plattformar och gemenskap – en analys av smart mobilitet utifrån delningsekonomins kännetecknande drag (Shared smart mobility? Access, platforms and community - an analysis of smart mobility based on the characteristic features of the sharing economy) K2 Working paper series 2020: 5

Papa, E., & Lauwers, D. (2015). Smart mobility: Opportunity or threat to innovate places and cities. In *20th international conference on urban planning and regional development in the information society (REAL CORP 2015)* (pp. 543–550). Ghent, Belgium.

Paulsson, A., & Mukhtar-Landgren, D. (2019). *Med algoritmen bakom ratten: Autonoma fordon och experimentell stadsplanering i Göteborg (With the algorithm behind the wheel: Autonomous vehicles and experimental urban planning in Gothenburg)* in Fronesis No. 64–65.

Rittel, H. W. J., & Webber, M. M. (1974). Wicked problems. *Man-made Futures*, *26*(1), 272–280.

Schuurman, D., & De Marez, L. (2012). Structuring user involvement in panel-based living labs. *Technology Innovation Management Review*, *2*(9), 31–38.

Sengers, F., Wieczorek, A., & Raven, R. (2019). Experimenting for sustainability transitions: A systematic literature review. *Technological Forecasting and Social Change, 145*, 153–164.

Smas, L., Schmitt, P., Perjo, L., & Tunström, M. (2016). Positioning urban labs – A new form of smart governance?. In *REAL CORP 2016–SMART ME UP! How to become and how to stay a Smart City, and does this improve quality of life? Proceedings of 21st International Conference on Urban Planning, Regional Development and Information Society*. (pp. 919–923). Hamburg, Germany: CORP–Competence Center of Urban and Regional Planning.

Steen, K., & van Bueren, E. (2017). The defining characteristics of urban living labs. *Technology Innovation Management Review*, *7*(7), 21–33.

Svara, J. H. (2006). Introduction: Politicians and adminis- trators in the political process—A review of themes and issues in the literature. *International Journal of Public Administration*, *29*(12), 953–976.

Van der Heijden, J. (2016). Experimental governance for low-carbon buildings and cities: Value and limits of local action networks. *Cities*, *53*, 1–7.

Vinnova. (2019). The next generation's travel and transport. Retrieved from https://www.vinnova.se/en/m/travel-and-transport/. Accessed on December 13, 2019.

Wiek, A., & Lang, D. J. (2016). Transformational sustainability research methodology. In H. Heinrichs, P. Martens, G. Michelsen G, & A. Wiek (Eds.), *Sustainability science* (pp. 31–41). Dordrecht: Springer.

Williams, J. (2016). Can low carbon city experiments transform the development regime?. *Futures*, *77*, 80–96.

Wolfram, M., Frantzeskaki, N., & Maschmeyer, S. (2016). Cities, systems and sustainability: Status and perspectives of research on urban transformations. *Current Opinion in Environmental Sustainability*, *22*, 18–25.

von Wirth, T., Fuenfschilling, L., Frantzeskaki, N., & Coenen, L. (2019). Impacts of urban living labs on sustainability transitions: Mechanisms and strategies for systemic change through experimentation. *European Planning Studies*, *27*(2), 229–257.

Voytenko, Y., McCormick, K., Evans, J., & Schliwa, G. (2016). Urban living labs for sustainability and low carbon cities in Europe: Towards a research agenda. *Journal of Cleaner Production*, *123*, 45–54.

Part III

Chapter 8

Smart Mobility as a Catalyst for Policy Change Towards Low Carbon Mobility?

Louise Reardon

ABSTRACT

With the plethora of smart mobility innovations, their applications, and their pace of change, it is easy to get distracted by what these innovations can (potentially) do, rather than what we want or need them to do, if we are to meet our societal goals. The focus of this chapter is therefore on the extent to which smart mobility can help create policy change towards the goal of low carbon mobility. The concept of policy is broken down into its component parts, to outline the relationship between policy goals and policy instruments, and identifies the key tools underpinning policy instruments. In turn, the chapter situates policy instruments within an understanding of policy change and triggers for policy change, arguing there are two key ways in which transformative change can occur; exogenously and endogenously. The chapter argues that the onset of smart mobility does not suggest an exogenous shock to the current policy system, in which smart mobility disrupts the authority and beliefs inherent within the current policy approach to mobility. Smart mobility therefore in and of itself is unlikely to lead to a radical policy shift towards low carbon. However, in understanding smart mobility innovations as policy instruments, it is possible to envisage smart mobility incrementally changing policy towards low carbon mobility, if opportunities for reflexivity and learning are embedded within local policy contexts.

Keywords: Policy change; low carbon mobility; smart mobility; learning; reflexive governance; incrementalism

Shaping Smart Mobility Futures:
Governance and Policy Instruments in times of Sustainability Transitions, 139–151

doi:10.1108/978-1-83982-650-420201008

Introduction

The aim of this chapter is to identify the ways in which smart mobility may enable transformative policy change towards low carbon mobility. Here, smart mobility is understood as a collection of technological innovations that have the potential to change the forms and ways in which people and goods move around. Smart mobility innovations therefore include autonomous vehicles that remove the need for a human driver; mobility as a Service (MaaS) that replaces a model of vehicle ownership to one of 'usership' through different membership packages; and 'intelligent' infrastructures capable of interacting with users and vehicles in real time to adapt services (Marsden & Reardon, 2018a, chapter 1, p. 3). With the plethora of smart mobility innovations, their applications, and their pace of change, it is easy to get distracted by what these innovations can (potentially) do, rather than what we want or need them to do, if we are to meet our societal goals.

The focus of this chapter is therefore on the extent to which smart mobility can help create policy change towards the goal of low carbon mobility. The transport sector accounts for 14% of global emissions (IPCC, 2014, p. 47), while in the UK transport accounts for 28% of total emissions (Committee on Climate Change (CCC), 2018, p. 147). Moreover, the transport sector is the fastest-growing consumer of fossil fuels, and fastest-growing source of CO_2 emissions globally (IEA, 2014). Therefore, as argued by the OECD, 'a radical and comprehensive approach to reducing emissions from transport is required' (OECD, 2015, p. 183). Movement towards 'low carbon mobility', in which there is a significant reduction in the level of carbon produced by mobility, is therefore a key governance challenge (Givoni & Banister, 2013) and one that arguably requires a radical shift in transport policy.

It is the focus on smart mobility as a potential enabler of *policy* change that distinguishes this chapter from other attempts to understand the implications of smart mobility. There has been significant focus on the extent to which smart mobility innovations can successfully challenge and change the existing socio-technical system of high carbon mobility (Markard, Raven, & Truffer, 2012). However, the role of governance and political activity has been overlooked to date (Marsden & Reardon, 2018a, chapter 1). This chapter is therefore underpinned by recognition of the significant role policy, and governmental activity can and do play in facilitating and constraining moves to low carbon mobility. To this end, the chapter has the following five sections.

The first section outlines what is meant by policy and policy instruments. The second section conceptualises policy change and argues that policy change can occur in two key ways; exogenously and endogenously. Section three explores the ways in which smart mobility has the potential to be an exogenous catalyst for policy change, while section four explores the ways in which smart mobility may act as an endogenous means towards policy change, positively affecting policy instruments to enable low carbon mobility. The chapter then concludes in section five. The chapter argues that based on existing evidence, smart mobility is not likely to affect significant policy change in an exogenous way. However, there is the potential for smart mobility to successfully affect policy incrementally through endogenous policy change, to the benefit of achieving low carbon mobility.

Policy and Policy Instruments

Policy is often a label given to a multiplicity of quite different and wide ranging government activities. It is therefore important to identify its different components and potential meanings. Howlett and Cashore (2009) argue that 'policy' consists of a 'complex regime of ends and means-related goals (more abstract), objectives (less abstract), and settings (least abstract)', as illustrated in Table 1 (Howlett & Cashore, 2009, pp. 38–39, see also Marsden & Reardon, 2017, p. 240). Goals are the ultimate ends that underpin policymaking. Objectives operationalise the goals, while settings specify what is required to put the objectives into practice in specific real-world situations. Settings can also be understood as context-specific targets. While goals, objectives, and settings relate to the ends and aims of policy, means and tools are also key components of policy. 'Instrument logics' refer to the norms that guide the enactment of policy; the 'mechanisms' refer to the types of instruments used to implement objectives (e.g. tax incentives); and 'calibrations' refer to the way an instrument is operationalised in practice; for example, whether voluntary or mandatory standards are imposed. Howlett and Cashore's (2009) taxonomy therefore provides a lens through which to disaggregate the ends and means of government activity and their different effects on and from, smart mobility.

Table 1. Policy Taxonomy.

		Policy Content		
Policy focus		*High Level Abstraction*	*Programme Level Operationalisation*	*Specific on-the-ground Measures*
	Policy ends or aims	GOALS:	OBJECTIVES:	SETTINGS:
		What general types of ideas govern policy	What does policy formally aim to address?	What are the specific on-the-ground requirements of policy?
		For example, environmental sustainability	For example, carbon levels from mobility	For example, reduce emissions by X per cent
	Policy means or tools	INSTRUMENT LOGIC:	MECHANISMS:	CALIBRATIONS:
		What general norms guide implementation preferences?	What specific types of instruments are utilised?	What are the specific ways/settings by which the instrument is used?
		For example, welfare maximisation	For example, spending on active travel infrastructure	For example, X km of cycle lane laid

Source: Adapted from Howlett and Cashore (2009, p. 39).

In Howlett and Cashore's (2009) parlance policy instruments are referred to as mechanisms, but the terms can be used interchangeably. Key here is the recognition that instruments are implemented to support policy objectives, and informed by a logic of implementation, for example, around welfare maximisation, that guide the suite of instruments deemed acceptable and appropriate to achieving the overarching goals of government activity. In turn, policy instruments can be calibrated and applied in different ways.

Hood (1983) (and later revised by Hood and Margetts, 2007) articulated what has come to be a seminal way of thinking about the role of policy instruments, categorising the policy tools at government's disposal. Hood and Margetts (2007) argue that government activity, which is essentially about controlling society, needs two key capabilities – the ability to 'detect' and the ability to 'effect'. Detectors refer to all the instruments government uses for consuming information, while effectors 'are all the tools government can use to try to make an impact on the world outside' (Hood & Margetts, 2007, p. 3). In turn, they argue that there are four basic resources – Nodality, Authority, Treasure and Organisation (NATO) – that governments have by virtue of being a government, which they can draw upon in order to detect and effect (see also Wallsten, Sørensen, Paulsson, and Hultén, 2020, Chapter 9 of this volume).

Nodality refers to 'the property of being in the middle of an information or social network' (Hood & Margetts, 2007, p. 5). Hood and Margetts (2007, p. 5) argue that government is likely to have nodality for one of three reasons. First, because they are the figurehead, up to which information is passed. Second, due to their unique position of being able to see, many different cases and thus build up a store of information not available to others. Third, because government often occupies a central place in the network; '... the Rome to which all roads lead' (Hood & Margetts, 2007, p. 5). The second resource, 'Authority', refers to the possession of legal or official power which gives the government the ability to 'demand, forbid, guarantee, [and] adjudicate' (p. 5). In turn, authority is one of the defining properties of government, although the 'source, base and level' of authority can vary widely (Hood & Margetts, 2007, p. 6). Treasure, the third resource, denotes the possession of a stock of money or other things with the money like property of 'fungibility' – the capacity to be freely exchanged (such as bonds, assets, and capital). The fourth resource, 'organisation' refers to the possession and arrangement of a stock of people (and their variety of skills), land, materials, equipment and technology underpinned by a bureaucracy.

Each of these four basic resources can be used in different ways, are interconnected to greater or lesser degrees, and are subject to different limitations. As Hood and Margetts (2007, p. 6) note, for example, the government's position as figurehead may equip it with a strategic position from which to deliver and collect information, 'spending' this resource through the messages it chooses to send and receive. However, its ability to do so will be limited by its credibility. Moreover, authority gives government the ability to legally determine a set of processes, 'spending' this through official authority, but which as a resource is limited by legal standing. The government's treasure will enable it to influence outsiders and purchase a range of other capabilities, such as information, but this will be subject to its solvency. While organisation gives government the physical ability to act directly, using its own resources. Government will 'spend' this resource through 'physical processing' of the different elements, while it's limiting factor is its capacity, largely influenced by its treasure.

This treatment of policy instruments is intentionally high level in order to be able to provide an outline of the core categorisations within which different policy instruments may fit and which, by the nature of what government is, do not change to any great extent over time. In the context of thinking about smart mobility, this point is particularly pertinent. As Hood and Margetts (2007, p. 14) highlight, there is a plethora of 'new' issues to which governments look to respond, and in particular where technology is involved, there is a desire to create 'new' policy instruments in order to address them. However, it should be recognised that '... many of these new instruments can be understood as old instruments in a new technological context'.

Catalysts for Policy Change

Having identified what is meant by policy and how the notion of policy instruments sits within that, it is important to understand the catalysts for policy change. This is because the focus here is on the extent to which smart mobility can help transform policy towards low carbon mobility. Hall (1993, p. 278) argues that policy change occurs as a result of 'social learning' in which policymakers 'attempt to adjust the goals or techniques of policy in response to past experience or new information'. Here, the policy variables of goals, instruments ('mechanisms' in Table 1) and settings ('calibrations' in Table 1) are important for determining the nature of change. For Hall (1993) 'first order' policy change occurs when experience and new information lead to the changing of the settings of policy instruments. 'Second order' change refers to a situation in which policy instruments and their settings change. While 'third order' change occurs when not only the policy instruments and their settings change, but also when the goals underpinning the rationale for the instruments changes. When the 'ends and aims' of policy in Howlett and Cashore's (2009) parlance, change. Third order change therefore amounts to paradigm change, the most radical transformation, but with it the most unlikely.

There are three key elements to this latter type of change (Bache & Reardon, 2016, pp. 28–29; Hall, 1993). First, sociological rather than scientific factors are likely to take precedence in triggering change. For example, the role of political judgement over expert opinion within institutional arrangements, and exogenous factors that may affect the power of one set of actors to impose their paradigm over others. Second, where the locus of authority rests over policy, with paradigm change likely to occur alongside shifts in where this authority resides. Third, experimentation and policy failure. The argument being that policymakers will attempt to respond to anomalies and challenges to the existing paradigm which will eventually undermine the coherence and precision of the existing approach, leading to policy failures which in turn, further undermine the existing paradigm.

Such theorising therefore posits three expectations. First, that policy processes are typically stable. Second, that policy change results from a breakdown of the institutionalised 'policy monopoly' of the existing way of doing and thinking about policy (Bache & Reardon, 2016, p. 30). And third, that transformative paradigm change occurs through an alteration of policy beliefs and membership due to external (societal) changes. However, Howlett and Cashore (2009) argue that more attention should be paid to small scale policy change happening within the policy-making system. For example, highlighting the possibility that small changes to policy settings over time can lead to the original goals of policy looking very different. They also stress the importance of cumulative change, and the importance of understanding policy developments that do not deviate from the status quo, and those that do move in new directions, to create cumulative change over time.

Bache and Reardon (2016, p. 157) in their research on the impact of wellbeing on policy, therefore identify two pathways to paradigm change. The first is an exogenous pathway, in which first and second order changes occur endogenously (through learning and experience of policy actors) but failure to respond to challenges in the underlying policy subsystem leads to the original paradigm, and the authority of its advocates, being undermined. With paradigm change then being driven by exogenous factors. The second is an endogenous pathway, in which first and second order change again occurs within the policy-making system, but when they 'a) successfully respond to challenges to the policy subsystem; and b) move in the same direction over a period of time, these changes can legitimise the ideas that inform them and strengthen the authority of their advocates' creating political momentum for change alongside the policy means (Bache & Reardon, 2016, pp. 157–158). In thinking about the role of smart mobility for determining policy change, and therefore affecting the goal of low carbon mobility, we can therefore think about its potential in two ways, one as an exogenous trigger for change, and second as an endogenous one. Both will be discussed in turn.

Smart Mobility as an Exogenous Shock?

To achieve low carbon mobility requires the decoupling of three key elements of the current (high-carbon) mobility system: economic growth, demand for transport, and emissions from transport (Givoni, 2013). The move to smart mobility has been heralded as one of the key ways to achieve this decoupling; reducing emissions, while not impeding demand, and at the same time supporting economic growth. For example, autonomous vehicles may require fewer safety features and therefore be smaller and lighter than current vehicles, making them better suited to electric power (Burns, 2013). Their self-driving nature also enables greater vehicle sharing as their need to be parked by an individual user is removed. Mobility as a Service also enables sharing and potentially greater uptake of public transport and active travel modes to overcome the last mile problem, meaning that mobility needs can be accommodated more efficiently; reducing idling and congestion and in turn reducing average travel times, which also has positive benefits for productivity and economic growth (Thomopoulos & Givoni, 2015, p. 13). Smart mobility, so its proponents argue, therefore has the potential to advance a green revolution in mobility.

Smart mobility constitutes an 'exogenous' factor in the sense that it involves a range of technological changes to mobility and potential behavioural changes within society, that lay outside of the policy arena. However for smart mobility to constitute an exogenous shock to the policy arena requires it to alter the locus of authority within the policy system and undermine the existing high carbon mobility paradigm upon which the current mobility system is based, subsequently challenging the beliefs and in turn basis on which current policy is made. However, current experience suggests the current locus of authority is not changing significantly. A recent edited collection by Marsden and Reardon (2018a, chapter 1) highlights a range of concerns about the transition to smart mobility reinforcing, rather than undermining, the existing mobility paradigm. Including the risk of a preference for conventional business models of private vehicle ownership, for example, in the move to autonomous and electric vehicles, and as a basis for MaaS (see also Schwanen, 2016). As Patterson et al. (2017, p. 10) note, actors who promote changes to the status quo (such as smart mobility innovators) '... do so from particular political perspectives, carrying with them a set of worldviews and values that influence their vision of what constitutes a desirable future'. Therefore 'Concerns relating to whose knowledge counts, what changes are necessary and desirable, and even what constitutes the end goal of transformation are all intensely political processes' (Patterson et al., 2017, p. 10). These political processes are yet to be worked through. In essence then, in the parlance of Howlett and Cashore (2009), it is not yet clear that the 'instrument logic', the norms underpinning the existing goals of transport policy and mobility, are being altered by the proliferation of smart mobility innovations.

Moreover, contrary to the prevailing discourse and optimism of what smart mobility can achieve for low carbon mobility, research has found that these green outcomes will not be achieved without government proactivity. Docherty, Marsden, and Anable (2018), for example, in their assessment of current smart mobility innovations, note how claims to environmental benefits from smart mobility (coming from greater system efficiency, for example) are based on models which require a proactive and unprecedented role of the state. For example, with smart mobility only demonstrating greater efficiency when there is one system manager, one or two forms of shared mobility that everyone will use, and an acceptance of sharing which is imposed by the system operator (Docherty et al., 2018, p. 119). Moreover, Pangbourne, Mladenović, Stead, and Milakis (2020, p. 9) note how MaaS may undermine public transport use, with 'impulsive' door to door journeys becoming 'incredibly convenient' through ride hail services, and the potential for increased discretionary trips to improve customer notions of value for money in paying monthly (rather than per trip, for example) for MaaS services. As Docherty et al. (2018, p. 114) therefore conclude:

> A failure to address both the short and longer-term governance issues risks locking the mobility system into transition paths which exacerbate rather than ameliorate the wider social and environmental problems that have challenged planners throughout the automobility transition.

Proactive governance of this kind, therefore requires recognition of the importance of the policy goal of climate change over and above other goals that are potentially facilitated by smart mobility but which may be negatively affected if climate change is prioritised. In relation to MaaS, for example, Pangbourne et al. (2020) identify economic growth as the primary driver of national government support for innovation in MaaS systems in multiple countries and for promoting them globally. The same can be said of autonomous vehicles in the UK, which are embedded in the UK's industrial strategy and economic growth ambitions (Hopkins & Schwannen, 2018). This focus on economic growth is compounded by the use of mobility and facilitation of travel demand as a 'solution' for achieving economic growth more generally, which in turn depoliticises its negative externalities or treats these as a secondary concern (Reardon & Marsden, 2020).

Smart mobility will therefore not be a catalyst for change towards the policy goal of low carbon mobility in and of itself, despite what the rhetoric of smart mobility innovators might suggest. One because it is not clear that the existing power relationships within the current policy system are being undermined or changed in a way that would facilitate low carbon mobility more readily. Two, because of the need for government to steer towards these outcomes and therefore recognise low carbon mobility as the highest level policy goal. In thinking though the lens of the exogenous model of change (Bache & Reardon, 2016), smart mobility may therefore be understood as currently being accommodated within the existing policy framework. Exogenous change may therefore be facilitated only if this accommodation of smart mobility systems leads to further undermining of the current mobility system and policy approach to it – exacerbating issues within the high carbon mobility system, which would then leave policymakers nowhere to hide in terms of the need for a different policy approach.

Smart Mobility as Policy Instrument(s) for Endogenous Change

While the conclusion of the previous section may have been negative in terms of the potential for smart mobility to have a positive transformative impact on low carbon mobility, smart mobility may have a positive impact endogenously. Simply put, the above conclusion is that smart mobility will only lead to an exogenous transformative change if its outcomes for low carbon mobility are so bad that they undermine the beliefs and norms upon which the existing system is based. However, as recognised by Marsden and Reardon (2018b, Chapter 9, p. 139) context matters in terms of smart mobility innovations and

> deciding who should act, how and at what scale, is crucial in setting the conditions in which new mobility systems can flourish, but in a way that promotes the goals of … governments and meets the needs of citizens ….

It is therefore possible that smart mobility could have a gradual, yet transformative effect towards low carbon mobility, through its role as a policy instrument or collection of instruments. The discussion in this section therefore focusses on the extent to which smart mobility not only affects existing policy instruments, but can also be an opportunity to alter policy instruments and their calibrations, these changes in turn creating a succession of incremental changes which over time transform policy towards low carbon mobility.

The transport studies literature highlights the need for 'integrated' governance structures for low carbon mobility (Buehler, Pucher, & Dümmler, 2019; Hickman & Banister, 2014; Rode & Cruz, 2018). The argument being that better regulatory and resource alignment between network institutions responsible for land-use, public transport, infrastructure and so on will mean improved ability to coordinate transport modes, restrict car use and therefore more easily enable travel demand management and facilitate low carbon mobility. Here then, the policy tool of 'organisation', as highlighted by Hood (1983) is key. However, smart mobility innovations, at least in the way they are currently being promoted, risk making such a synergy of organisational approach harder to achieve. In the UK, for example, more institutional silos are being created around transport rather than less. With regards to autonomous vehicles, for example, a Centre for Connected and Autonomous Vehicles has been jointly created by the Department for Transport and the Department for Business, Energy, and Industrial Strategy. While on the surface this institutional arrangement has the potential to foster cross-department collaboration, it also risks isolating the development of this mode from systemic issues such as climate change (Hopkins & Schwanen, 2018).

However smart mobility, in its application, has the potential to redefine the ways in which the instrument of 'organisation' is applied. Dunleavy, Margetts, Bastow, and Tinkler (2006, p. 467), for example, heralded in the era of 'digital era governance' over a decade ago, arguing that the internet and information technology systems have the potential to remove bureaucratic silos, 'reintegrating functions into the governmental sphere, adopting holistic and needs-oriented structures and progressing digitalization of administrative processes'. Smart mobility, with the technological advances they bring, could do the same. While not replacing the need for coherent organisational arrangements within which decisions are made, smart mobility has the potential to aid the integration of systems and the development of more 'open-minded' governance structures, in which data collection enabled through these new forms of mobility is used to better understand the context in which low carbon mobility is required, for example, land-use, journey patterns, and mode use, enabling more informed policy interventions (Meijer & Bolivar, 2016), in turn strengthening rather than reducing 'organisation' as a policy instrument. However, such organisational benefits not only require integrated decision-making structures in order for this data to be used effectively, but also collaborative and 'smart governance' alongside it, in which governments and the private sector work together to share access to information and work together to address societal challenges such as climate change (Meijer & Bolivar, 2016).

Related to the above point on integration as being a key enabling factor for a move to low carbon mobility, at the local level the existence of Metropolitan

Strategic Authorities has been heralded as important enablers in cities, where vehicle miles travelled have been reduced and increased use of public transport has been achieved (Buehler et al., 2019; Rode, Heeckt, & da Cruz, 2019). The idea here is that this organisational arrangement is a more effective way of pooling government authority, treasure and nodality across a geographic area, which in turn means that greater connectivity and more integration across modes can be achieved. However, smart mobility has the potential to make the geographical legislative boundaries of state authorities less meaningful or pertinent for the smooth running of the system. For example, with the onset of MaaS, transport authorities are less likely to be default system aggregators (the nodal figurehead and clearing house) for service information, while the onset of autonomous vehicles, e-scooters and the like, may reduce the demand and therefore significance of modes commonly controlled by government organisations. Smart mobility may therefore require the principles or notion of Metropolitan Strategic Authorities to be reimagined, with their emphasis shifting towards the coordination of data and the setting of regulatory arrangements and parameters for system activity, rather than delivery of a service. In turn, utilising the resource of authority to underpin this policy instrument, rather than that of physical 'organisation' per se; essentially governing governance. The literature points to the importance of support and facilitation of networks as important tools for such meta-governance over and above more traditional hierarchical, top-down mechanisms of control (Sørensen, 2006; Torfing, Peters, Pierre, & Sørensen, 2012).

In terms of treasure, Rode et al. (2019) identify infrastructure budget reallocation and land value capture as key ways to foster compact and connected urban growth and in turn foster low carbon emission environments. In particular, budget allocation is identified as 'fundamental', with infrastructure playing a key role in enabling or undermining different modes of travel and in turn fostering more or less sustainable modes of transport. It is therefore argued that significant investment should be put towards public transport and active travel. However, the amount of treasure may be significantly undermined by a move to smart mobility if a proactive approach is not taken. For example, as discussed earlier, autonomous vehicles and MaaS may foster an individualised approach to transport which undermines the viability of public transport, which may therefore require higher subsidy and therefore become more expensive to run and in turn make it harder to justify expansion of the network. However, smart mobility does not remove the state's ability to gain treasure through its authority and could create new revenue streams for this end. For example, a levy could be imposed on smart mobility operators that could be invested back in to public transport as a new means of raising 'treasure'. Moreover, while there is very little appetite within the UK, for example, to intervene in or impede commercial operations (Docherty, 2018), smart mobility operators could be incentivised through the tax system to act more sustainably. For example, with tax breaks depending on how clean their vehicle fleet is, or on whether they bundle active travel within their MaaS systems.

In terms of authority, it is argued that parking standards are an important lever for creating low carbon environments (Rode et al., 2019). For example, being able to impose maximum requirements, reducing incentives such as free parking and imposing restrictions on where to park can create more liveable environments and also encourage modal shift to more sustainable modes. Authority, in this regard, has the potential to be significantly undermined by smart mobility. Autonomous vehicles, for example, may not need to be parked (at least not in central business areas) and indeed, the loss of revenue from parking may reduce the amount of 'treasure' at the government's disposal, in turn compounding the issues discussed above. However, it can be argued that parking is not currently used as a policy lever as much as it could or should be for enabling low carbon mobility. For example, in the UK the workplace parking levy is only currently utilised by one local authority that has the power to do so (Transport for London (TfL), 2019). Given that the transition to smart mobility, in particular innovations like autonomous vehicles are likely to be gradual it is possible to argue that a move to these technologies may make the application of parking standards more politically palatable (as there may be less demand to park), but still have an influence on travel behaviour. Moreover, smart mobility may enable other policy instruments derived from authority and nodality to become technically (if not politically) easier to apply, for example, congestion charging.

In terms of smart mobility facilitating endogenous change towards low carbon mobility, each of the potentially positive and negative consequences of policy instrument change discussed above may be possible. However as highlighted earlier, a positive outcome will be dependent on the context and governance approach taken in each locality and towards each different form of smart mobility (Marsden & Reardon, 2018b, Chapter 9). However, as emphasised by Hall (1993), change is predicated on the ability to learn from previous policy experience and policy failure, altering policy instruments and their settings, in response to the lessons these experiences provide. With the onset of smart mobility innovations, there is therefore the need for more 'reflexive' forms of governance that more readily enable this learning to take place (Hopkins & Schwannen, 2018). Reflexive governance evokes the need for a process of collaboration across stakeholders, in which debate, research, and assessment are used to monitor and evaluate interventions to foster learning. However, such processes can be hard to embed. As Hopkins and Schwannen (2018, p. 79) note in relation to autonomous vehicle development in the UK '... there appears to be little space for critical interventions, big picture thinking, reflexivity or indeed failure'. They argue that this can be put down to, at least in part, '... the way in which the smart mobility transition is framed in the context of neo-liberal ideals and dependence on private sector interests' (p.79). Such observations highlight the extent to which goals, and the norms underpinning them, as outlined by Howlett and Cashore (2009), influence objectives and the subsequent application of policy instruments, such as those embedded within smart mobility innovations. However, this is not to say that reflexivity and positive incrementalism towards low carbon mobility are not possible through smart innovations, but rather that it is not guaranteed and will therefore require open contestation about the impacts of smart mobility on the suite of policy instruments available for working towards low carbon outcomes.

Conclusion

This chapter has sought to outline the potential ways in which smart mobility may affect policy change towards low carbon mobility. The concept of policy has been broken down into its component parts, to outline the relationship between policy goals and policy instruments, and identified the key tools underpinning policy instruments through articulation of the NATO framework. In turn, the chapter has situated policy instruments within an understanding of policy change and triggers for policy change, arguing there are two key ways in which transformative change can occur; exogenously and endogenously. It has been argued that the onset of smart mobility does not suggest an exogenous shock to the current policy system, in which smart mobility disrupts the authority and beliefs inherent within the current system and therefore facilitates a radical policy shift towards low carbon mobility. However, in understanding smart mobility innovations as policy instruments, it is possible to envisage smart mobility incrementally changing policy towards low carbon mobility if opportunities for reflexivity and learning are embedded within local policy contexts.

References

Bache, I., & Reardon, L. (2016). *The politics and policy of wellbeing: Understanding the rise and significance of a New Agenda*. London: Edward Elgar.

Buehler, R., Pucher, J., & Dümmler, O. (2019). Verkehrsverbund: The evolution and spread of fully integrated regional public transport in Germany, Austria, and Switzerland. *International Journal of Sustainable Transportation, 13*(1), 36–50.

Burns, L. (2013). A vision of our future, *Nature, 497*, 81-182.

Committee on Climate Change (CCC). (2018). *Reducing UK emissions*. London: Committee on Climate Change.

Docherty, I., Marsden, G., & Anable, J. (2018). The governance of smart mobility, *Transportation Research Part A. 115*, 114–125.

Dunleavy, P., Margetts, H., Bastow, S., & Tinkler, J. (2006). New public management is dead: Long live digital-era governance. *Journal of Public Administration Research and Theory: J-PART, 16*(3), 467–494.

Givoni, M., & Banister, D. (2013). *Moving towards low carbon mobility*. Cheltenham: Edward Elgar.

Hall, P. (1993). Policy paradigms, social learning, and the state: The case of economic policymaking in Britain. *Comparative Politics, 25*(3), 275–296.

Hickman, R., & Banister, D. (2014). *Transport, climate change and the city*. Routledge: Abingdon.

Hood, C. (1983). *The tools of government*. London: Springer.

Hood, C., & Margetts, H. Z. (2007). *The tools of government in the digital age*. London: Palgrave Macmillan.

Hopkins, D., & Schwanen, T. (2018). Governing the race to automation. In G. Marsden & L. Reardon (Eds.), *Governance of the smart mobility transition* (pp. 65–84). London: Emerald.

Howlett, M., & Cashore, B. (2009). The dependent variable problem in the study of policy change: Understanding policy change as a methodological problem. *Journal of Comparative Policy Analysis: Research and Practice, 11*(1), 33–46.

IEA. (2014). *World energy investment outlook: special report*. Paris: OECD/IEA Publishing.

IPCC. (2014). Climate Change 2014 Synthesis Report. *Contribution of Working Groups I, II and III to the Fifth Assessment Report of the Intergovernmental Panel on Climate Change* IPCC, Geneva, Switzerland.

Givoni, M. (2013). Alternative pathways to low carbon mobility. In M. Givoni & D. Banister (Eds.), *Moving towards low carbon mobility*. Cheltenham: Edward Elgar, 209–230.

Markard, J., Raven, R., & Truffer, B. (2012). Sustainability transitions: An emerging field of research and its prospects. *Research Policy*, *41*, 955–967.

Marsden, G., & Reardon, L. (2017). Questions of governance: Rethinking the study of transportation policy. *Transportation Research Part A: Policy and Practice*, *101*, 238–251.

Marsden, G., & Reardon, L. (2018a). Introduction. In G. Marsden & L. Reardon (Eds.), *Governance of the smart mobility transition* (pp. 1–15). London: Emerald.

Marsden, G., & Reardon, L. (2018b). Does governance matter? An international scenarios exercise. In G. Marsden & L. Reardon (Eds.), *Governance of the smart mobility transition* (pp. 139–151). London: Emerald.

Meijer, A., & Bolivar, M. (2016). Governing the smart city: A review of the literature on smart urban governance. *International Review of Administrative Sciences*, *82*(2), 392–408.

OECD. (2015). *Aligning policies for a low-carbon economy*. Paris: OECD Publishing.

Pangbourne, K., Mladenović, M, Stead, D., & Milakis, D. (2020). Questioning mobility as a service: Unanticipated implications for society and governance. *Transportation Research Part A*, *131*, 35–49. https://doi.org/10.1016/j.tra.2019.09.033

Patterson, J., Schulz, K., Vervoort, J., van der Hel, S., Widerberg, O., Adler, C., … Barau, A. (2017). Exploring the governance and politics of transformations towards sustainability. *Environmental Innovation and Societal Transitions*, *24*, 1–16. http://dx.doi.org/10.1016/j.eist.2016.09.001

Reardon, L., & Marsden, G. (2020). Exploring the role of the state in the depoliticisation of UK transport policy. *Policy & Politics*, *48*(2), 223–240.

Rode, P., & da Cruz, N. (2018). Governing urban accessibility: Moving beyond transport and mobility. *Applied Mobilities*, *3*(1), 8–33.

Rode, P., Heeckt, C., & da Cruz, N. (2019). *National transport policy and cities: Key policy interventions to drive compact and connected urban growth*. London: LSE Cities.

Schwanen, T. (2016). Rethinking resilience as capacity to endure. *City*, *20*(1), 152–160.

Sørensen, E. (2006). Metagovernance: The changing role of politicians in processes of democratic governance. *The American Review of Public Administration*, *36*(1), 98–114.\

Thomopoulos, N., & Givoni, M. (2015). The autonomous car—a blessing or a curse for the future of low carbon mobility? An exploration of likely vs. desirable outcomes, *European Journal of Futures Research*, *3*(14), 1–14.

Torfing, J., Peters, B. G., Pierre, J., & Sørensen, E. (2012). *Interactive governance: Advancing the paradigm*. Oxford: Oxford University Press.

Transport for London (TfL). (2019). Workplace parking levies: Frequently asked questions. Retrieved from http://content.tfl.gov.uk/frequently-asked-questions-on-wpls.pdf. Accessed on November 2019.

Wallsten, A., Sørensen, C. H., Paulsson, A., & Hultén, J. (2020). The governing capacity of policy instruments in three different smart mobility futures. In A. Paulsson & C. H. Sørensen (Eds.), *Shaping smart mobility futures. Governance and policy instruments in times of sustainability transitions* (pp. 153–168). Bingley: Emerald.

Chapter 9

Is Governing Capacity Undermined? Policy Instruments in Smart Mobility Futures

Anna Wallsten, Claus Hedegaard Sørensen, Alexander Paulsson and John Hultén

ABSTRACT

The aim of this chapter is to analyse how the governing capacity of current policy instruments might be affected in futures of smart mobility. In order to explore this issue, the authors make use of the so-called NATO (*nodality, authority, treasure, organisation*) framework for analysing two contrasting scenarios. The analyses show that the overall governing capacity of many of the policy instruments is strengthened or maintained in both of the scenarios. However, the governing capacity of some policy instruments is reduced, and some seem to need calibration, not least because authorities' access to and control over data are under question. Future governing capacity hinges on access to data, although all resources are, in one way or another, affected.

Keywords: Policy instruments; governing capacity; smart mobility futures; NATO framework; scenarios; governance

Introduction

Smart mobility is often defined as the autonomous, shared and integrated features of future transportation facilitated by digitalisation and electrification. It is currently widely debated in academic as well as industry settings, and an increasing amount of literature stresses the need of governance to ensure that the emerging technologies contribute towards societal objectives (Finger & Audouin, 2018; Marsden & Reardon, 2018). Unless governance is included in the equation, there

Shaping Smart Mobility Futures:
Governance and Policy Instruments in times of Sustainability Transitions, 153–168

doi:10.1108/978-1-83982-650-420201009

is a risk that the transition to smart mobility may lead to difficulties in achieving societal objectives. In the many modelling studies published (for an overview, see Stead & Vaddadi, 2019), both advantages and disadvantages are evident. On the one hand, the modelling studies show that improved traffic safety and a more efficient transport system can be materialised by harnessing the benefits coming from the new technological advancements. On the other hand, many of these studies show that automated cars will lead to more traffic, contribute to urban congestion and challenge the high ambitions of reducing greenhouse gas emissions as well as the ambitions of making mobility accessible for all.

While a demand for governance is stressed in several recent publications, literature discussing the actual policy instruments available and necessary to govern a transition to smart mobility is less discussed. Yet, grey papers and reports from government agencies and consultancies, scrutinising the need of policy instruments to prevent negative consequences of certain future scenarios, may be found along with analyses of possible new policy instruments made available by the new emerging technologies (e.g. Swedish Transport Administration, 2018; Transport Analysis, 2017, 2019). Missing among these publications, however, are analyses of the strengths and weaknesses of current policy instruments in the transport sector in different futures of smart mobility. As an example, it is imaginable that parking restrictions and parking fees in cities might not be an adequate and strong policy instrument in a future where self-driving vehicles can continue circulating or park outside of the city.

All in all, this suggests that the *governing capacity* of various currently deployed policy instruments might be affected in futures of smart mobility. The terms 'governing capacity', 'institutional capacity' (Cars, Healy, Madanipour, & Magalaes, 2002) or 'policy capacity' are usually applied when analysing a nation-state's possibility to control either a certain geographical area, or a policy area (Field, 2016; Painter & Pierre, 2005). Building upon Max Weber's notion that states may only be defined by their methods or means, not their objectives or substantive purposes, Offe (2015) has suggested that governing capacity must be understood alongside its opposite, ungovernability. Offe (2015, p 79) argues that ungovernability is a result of creeping inaction and incapacity. When public institutions allow certain kinds of problems and conflicts to emerge, the same institutions might later turn out to be incapable of governing and processing those problems (partly because the problems have grown in proportions and governing them would require massive efforts by the public institutions). The recent and ongoing financial and environmental crises could be understood as examples of this. In fact, Offe terms these policy areas as ungovernable because they pose such massive challenges for public institutions – on so many scales and on such different time horizons. Yet, ungovernability also has a normative component, as it reflects which policy areas that states are expected to be able to govern. This normative component seems rather uncontroversial, though, as many, if not most, people – across the political spectrum – agree that if a state is lacking the ability to 'make and enforce laws, provide basic services, or resolve major conflicts through adequate institutional means', it is also lacking in terms of governability and governing capacity (Offe, 2015, p 79).

In this chapter, the term governing capacity is used in analyses of policy instruments. By 'governing capacity', we mean the ability of an instrument to establish the anticipated or wished effect. As will be explained below, governing capacity is conceptualised as the four resources identified in the NATO (nodality, authority, treasure, organisation) framework (Hood & Margetts, 2007). While there have been calls for governance of smart mobility in the literature, the topic of governing capacity has not been debated, though it is highly relevant for political-administrative authorities as well as for scholars to consider. Going back to Offe's argument, if a state is lacking governing capacity, this can be interpreted as a sign that states should not govern certain policy areas at all. Rather than building capacity and strengthening the policy instruments, the lack of governing capacity could then lead to two conclusions: either that there should be no conscious governance and development of policy instruments or that governance should be reinforced to guarantee future governing capacity.

This chapter refers to empirical studies carried out in Sweden, and is adapted from a report written in Swedish (Wallsten et al., 2019) focussing on current policy instruments deployed at the national level. The research question guiding this study is therefore: how is the governing capacity of current policy instruments affected in smart mobility futures? Smart mobility is defined here as mobility characterised by some degree of autonomous and connected vehicles, as well as shared mobility, though the mix of these characteristics differ between the two scenarios, which we will introduce later.

This brief introduction is followed by three sections. In the second section, the analytical approach of the chapter is described. We will explain the NATO framework for understanding policy instruments based on four resources, which is applied to structure the analysis. The third section is divided into two sub-sections, each taking its point of departure in two different futures of smart mobility, namely *Individualism* and *Sharing Economy*. In each sub-section, the scenario is presented, followed by an analysis of how the scenario will affect the governing capacity of policy instruments related to each of the four NATO resources. The fourth section wraps up the analysis and presents conclusions, and we further discuss what the implications of the analyses are for state authorities in and outside Sweden.

Analytical Approach

The future is highly uncertain, but through various methods, such as forecasting, backcasting, and explorative methods, previous studies have tried to identify possible lines of development in relation to smart mobility futures (e.g. Kristoffersson, Pernestål Brenden, & Mattson, 2017; Stead & Vaddadi, 2019). In this study, the point of departure is two scenarios of smart mobility futures developed by Pernestål and Almlöf (2019). In order to categorise and analyse how the state's governing capacity might be affected in these two scenarios, we have used the NATO framework proposed by Hood and Margetts (2007). We assess policy instruments used on a national level in the transport sector in Sweden. Despite the Swedish focus, we believe the conclusions are of greater relevance and that the results could be applicable to other national contexts as well.

We have used a variety of qualitative methods. Through document analyses, workshops and a questionnaire to key players employed in state authorities, we compiled a comprehensive overview of current policy instruments. We then analysed how the governing capacity of these instruments could be affected in the two scenarios. In the analyses, we have been inspired by literature on smart mobility, but since this is an emerging field our assessments and results should be regarded as preliminary and primarily as an early attempt to explore the state's governing capacity in smart mobility futures.

Categorisations of Current Policy Instruments

Hood and Margetts (2007) argue that governments shape people's lives and societies through various policy instruments. They suggest that policy instruments can be used to handle specific problems, reused to handle new problems, combined in different ways and adjusted to fit a certain context (Hood & Margetts, 2007). Using this definition as a basis, Hood and Margetts (2007) outline four categories of resources available to governments presented in a NATO framework. In this section, we show how we have operationalised this NATO framework to fit analysis of state policy instruments within the field of transport.

Nodality Nodality refers to the characteristic of being at the centre. A node is the place where several information channels are crossed and by being in this strategic hub, the state can effectively obtain and disseminate information. Through its nodality, the state becomes central, visible and interconnected with other leading actors (Hood & Margetts, 2007, chapter 2). When applying these conceptualisations on the transport system, we find three categories of policy instruments. The state can engage with *Mutual dialogue in social networks*, through which it can stimulate contacts between actors, initiate joint projects, keep up to date on innovations, research results, etc. Furthermore, the state can make use of *Digitalised data* that encompass all the information the state transmits and assembles via technical information channels such as intelligent transport systems (ITS), large databases, etc. *Unilateral state communication* entails the various ways through which the state disseminates information to individual citizens, groups or society at large, such information can, for example, be transmitted through social media, announcements, printed brochures, websites, etc.

Authority The state's authority is based on its ability to require, prohibit, guarantee or permit certain actions. These legally based forms of instruments can be used when the state wants to be sure of achieving a certain effect. At the same time, authority is nothing more than a symbolic exercise of power and is highly dependent on the state being respected and accepted (Hood & Margetts, 2007, chapter 3). The regulations governing the transport systems are not isolated to national borders, rather, they follow international agreements and the municipal and regional levels also have considerable authority. With that in mind, the analysis in this chapter is confined to instruments on a national level. We have found four categories through which the state expresses its authority within the transport system: *Taxes and fees,* which includes vehicle tax, congestion tax, tolls, fuel tax, etc. *Regulations for traffic, drivers and vehicles,* which contains the regulations for conditions at a national level, for example, license requirements, speed limit requirements, regulations targeting vehicles. Another category is *Regulations for land use and planning,* which contains the regulations through which the state governs the planning of societies. Finally, *Data management regulations* are an increasingly important category within the transport system as it contains regulations governing how organisations and companies can possess and process data.

Treasure The state also has the resource of interchangeable property, which primarily entails money. This resource can be used to obtain information, to buy goods or services, to support specific groups as a reward for encouraging certain activities or to get actors to produce or promise something through contracts (Hood & Margetts, 2007, chapter 4). When it comes to the transport system, we have found three categories of ways that the state governs through its treasure. *Infrastructure investments* which contain the financial resources that the state use in order to influence, for example, road and rail infrastructures. The category *R&D efforts* entails the financial support that the state can offer for various forms of research, development and demonstration activities. Finally, the category *Subsidies* entails the money used by the state to initiate certain developments or different modes of transport, for example, public transport.

Organisation This resource includes the state's competence and capacity. It refers to the composition of buildings, equipment and individuals that the state can make use of (Hood & Margetts, 2007, chapter 5). We have found four different categories within this resource that can apply to the transport system. To begin with, *The scope and structure of the state* involves how the state is organised in different authorities including their respective mandates and roles. *Employees and their competencies* encompass all the people who work at authorities including their knowledge and skills. The resource organisation also consists of the state's *Existing infrastructure* including buildings, roads, railways, etc.

All in all, the state uses a variety of different policy instruments to achieve transport policy objectives, which are based on four different resources. These resources are not easily separable, rather they reinforce each other, and individual instruments often act as a hybrid of different resources (Hood & Margetts, 2007). The categorisations upon which we base the upcoming analysis are summarised in Table 1.

Governing Capacity in Smart Mobility Futures

In this section, we use the NATO framework to analyse how the state's governing capacity might be affected in two scenarios of smart mobility futures. These scenarios were developed with the intention to show the range of possible alternative futures (Pernestål & Almlöf, 2019). This means that they should not be read as realistic futures, but rather as scenarios providing 'food for thoughts'; hence they can be interpreted as slightly provocative and extreme. The analysis does not

Table 1. Operationalisation of the Four NATO Resources.

NODALITY	AUTHORITY	TREASURE	ORGANISATION
– *Mutual dialogue in social networks*	– *Taxes and fees*	– *Infrastructure investments*	– *The scope and structure of the state*
– *Digitalised data*	– *Regulations for traffic, drivers and vehicles*	– *R&D efforts*	– *Employees and their competencies*
– *Unilateral state communication*	– *Regulations for land use and planning*	– *Subsidies*	– *Existing infrastructure*
	– *Data management regulations*		

take into account what kinds of policies that have led up to each scenario; rather we start off in the suggested futures and discuss the consequences each scenario might have on the state's governing capacity.

Individualism: The Transport System Still Centres Around the Private Car

In this scenario, the car park has grown significantly. Car traffic has increased by more than 80% compared to 2019. Increasing numbers of cars have resulted in growing queues, and parking space continues to be a challenge in urban areas. All new cars are electric, and they make up 70% of the car park. Although the cars are not completely self-driving (level 3), new cars have a high degree of self-driving (level 4) and on major roads there are infrastructures that help the cars reach full automation. This means that the driver can use the travel time for work or to rest which has enlarged the labour market regions. The long trend of urbanisation has consequently been broken, and citizens have moved out of the cities to residential towns or smaller communities in rural areas.

As the scenario title shows, this future is characterised by an increased degree of individualism. Integrity is thus important; people are reluctant to share personal information and they value riding in their own cars. A high degree of individualism has had a strong negative effect on public transportation. The demand for and the level of service of public transport has decreased, with subsequently increased social inequalities between those who have access to a private car and those who do not.

Finally, the scenario suggests an altered state role and mandate. The transport system is no longer proactively governed, instead, the state sees it as its task to control market actors and ensure the integrity of citizens and companies. Market actors, for example, engage with collecting and analysing data on various traffic patterns.

Increased Competition for Nodality. In this scenario, the state's capacity to govern through *mutual dialogue in social networks* will likely not change substantially compared to today; nevertheless, there is a chance that the composition of network members could be altered due to a more complex transport system. It is reasonable to expect that vehicle manufacturers, platform, and IT companies will be leading players in the networks, but that state actors still will have a position that enables them to make use of the networks to influence development.

The scenario proposes a technical development with connected vehicles and intelligent infrastructure which requires a constant flow of information. *Digitalised data* will become an essential asset for managing, forecasting and modelling traffic flows. However, the scenario rests upon assumptions of individuals' and organisations' unwillingness to share data. Withholding information can be a strategy to combat current competitors and obstruct the entrance of new market players (Transport Analysis, 2019). This is because data can be understood as the most valuable commodity on which the entire transition towards smart mobility is based, access to data provides power through which the market can be controlled (Docherty, Marsden, & Anable, 2018). Based on such expectations, we find it likely that the state's governing capacity for digitalised data might decrease in this scenario. The capacity of *unilateral state communication* might also decrease. This is mainly because it will not be considered a natural part of the state's commitment to inform citizens of what is 'right and wrong'. A reduced authority makes it reasonable to expect that the state becomes less visible as a source of information to the public (Hood & Margetts, 2007, p. 46).

All in all, the scenario represents a development that weakens the state's ability to govern through nodality. This is mainly explained by the widespread reluctance to share data, a more restrained view on the state's role and mandate and an increased competition for nodality from other actors.

Fees and Legal Compliance Increase Capacity of Authority. The scenario presents a future with more car traffic which means that the governing capacity for *taxes and fees* related to ownership and use of cars generally increases. The design of the vehicle tax will continue to influence consumers' purchases of cars in the same way as today, but with a significantly larger car fleet the overall effect of changes in the vehicle tax will be greater than today. The same correlation is also valid for other taxes such as congestion taxes and tolls. On the other hand, the capacity of the fuel tax, in its current design, will be significantly reduced in a future where electric cars make up 70% of the fleet (cf. Lindberg & Fridstrøm, 2015).

The governing capacity of *regulations for traffic, drivers and vehicles* is also likely to increase in this scenario. Digitalisation of the transport system involves new technologies that simplify the process of ensuring compliance with prevailing rules. Technological solutions including geo-fencing can force vehicles to adhere to speed limits, safe distance to other cars, winter tire regulations, etc. However, since this scenario stresses integrity issues, it is reasonable to expect the design solutions to be limited if the protection of personal data becomes more important than being able to control the transport system. For many other rules, such as driver's licenses, driver's competence, taxi and vehicle legislation, the steering capacity will presumably be comparable with today, as long as fully self-driving cars are not available, as this scenario suggests.

When it comes to *regulations for land use and planning*, we observe that the conditions will be altered by a more widespread geographical spread of where people spend their time. Nevertheless, we find that the concrete policy instruments in the form of, for example, the Planning and Building Act will still be important and expect that the governing capacity for such instruments will remain at today's level. Finally, with more data to handle, the governing capacity for *data management regulations* will increase. Self-driving vehicles at level 3 can be expected to deliver significantly larger amounts of data than is the case today (Bjelfvenstam, 2018; Swedish Transport Administration, 2018). At the same time, similar to the reasoning presented above, we assume that the state's regulation of data will focus on protecting personal integrity rather than using data to control the transport system. Despite increased amount of data, we thus find it likely that the governing capacity for data management regulations will be reduced in a future where the right to privacy is highly valued.

All in all, the governing capacity for authority is expected to increase in this scenario compared to today; this is due to increased traffic flows of private cars leading to enhanced governing capacity for taxes and fees, and increased regulatory compliance made possible by increased digitalisation.

The Need for Road Investments Increases the Capacity of Treasure. In this scenario, there is no indication that *infrastructure investments* would be less important than today. However, different types of infrastructure investments will likely have different levels of impact. Above all, increasing car traffic means increased capacity for measures in the road infrastructure. Through new investments and maintenance of state roads, and through contributions to municipal and private roads, the state can influence the development of road traffic. The purpose of these initiatives may be to uphold the increased private car flows that the scenario entails, or to limit the negative effects of such a development.

We expect the governing capacity of *R&D efforts* to likely increase compared with today. Demonstration projects have, for example, played an important role for the development of automated vehicles (Hopkins & Schwanen, 2018), and we assume that the technological development on which the scenario is based will bring new challenges in need of new innovations. At the same time, we know from history that the greatest potential for influence exists in the early stages of technological development, and by the time of the scenario, the potential may have weakened. We also want to stress that technology development is by nature global, making an individual state's ability to influence limited but not insignificant.

As for *subsidies*, the scenario does not suggest any obvious change in governing capacity, but due to increased market-oriented tendencies, the state might be less prepared to provide subsidies. If there is a political will, the state can subsidise transport services, for example, given that the subsidies are compatible with European regulations.

All in all, we expect that the governing capacity of treasure increases in this scenario, especially when it comes to investments in road infrastructure and R&D efforts.

A Reactive Organisation with Decreased Capacity. Due to far-reaching individualism and a reduced state mandate in the transport system, we find it likely that the governing capacity of *the scope and structure of the state* will change. The scenario presents a more reactive state, and we assume that it will primarily respond to market failures and operate through other actors rather than through clear-cut directives. The state might function as a prescriber and supervisor of various standards, as a developer of guidelines or a coach for market actors and individuals in terms of how they best can contribute to transport policy objectives. The governing capacity of *employees and their competences* will likely follow a similar trend. We assume that state employees will be given changed tasks and need other skills, but with a reduced mandate in the transport system the number of employees would probably also be decreased, resulting in a reduced governing capacity.

The *existing infrastructure* influences how individuals travel. The scenario presents a future with more dispersed travelling patterns and increased car traffic, which will likely increase the importance of infrastructural arrangements. It is also possible that the state can make use of digitalisation to obtain more updated and detailed information on the infrastructures' conditions and statuses (Swedish Transport Administration, 2017), enhancing the possibility of a more efficient usage. Based on these assumptions, we expect the governing capacity of existing infrastructures to increase in this scenario.

In summary, we assess that the governing capacity of the resource organisation will decrease in this scenario compared to today. This is mainly due to a reduced state mandate in the transport system, which is reflected in the state adopting a reactive role to govern the transport system through other actors.

Sharing Economy: The Breakthrough of Shared Mobility

In this scenario, the urban population has increased at the expense of a constantly decreasing number of inhabitants living in rural areas. Cities are the epicentre of development, not only do they attract new residents, new smart urban services are also constantly emerging. This scenario is also characterised by a widespread positive attitude towards freely sharing data both among citizens and companies. These sharing practices can also be seen in the transport sector, and people generally do not own vehicles for transportation. Instead, they use the closest shared scooter, shared bike, shared car or buses. Most people in the cities fulfil their mobility needs through a MaaS subscription.

Increased sharing has diminished the number of kilometres driven in the country. Electric cars make up 70% of the car park, all cars have advanced driver support (level 3) and new cars have a high degree of self-driving (level 4). The technological achievements within public transport follow similar tendencies as driverless buses have been introduced. Almost all vehicles are connected, which is generally perceived as unproblematic due to the willingness to share data. However, the accessibility for urban and rural citizens is increasingly unequal. In the cities, the development of shared mobility services is driven by profit-seeking companies who have not found convincing business cases in less populated areas. Shared initiatives driven by the rural civil society are common, but in general, the car remains a necessary transport mode in rural areas.

Finally, the state's commitment to govern the transport system has gradually increased and the state takes on a proactive role in steering towards the transport policy objectives. Due to this development, the presence of the state around the country has increased.

Nodality is Favoured by Positive Attitudes Towards Sharing Information. As in the previous scenario, it is likely that *mutual dialogue in social networks* will be affected in terms of altered network members and subject areas discussed, for example, MaaS providers and larger cities would possibly possess more pivotal network roles. A willingness to share information combined with an acceptance of an enhanced state mandate in the transport system makes it reasonable to expect that social networks will continue to be a valuable asset for the state in this scenario.

Given that the state has a leading role in the transport system, we assume that state actors will have access to essential digital infrastructure and this combined with a willingness to share information, we expect the state to have increased opportunities to govern through *digitalised data*. We assume that the state could streamline its planning and control of traffic flows through real-time data on traffic volume on roads, advanced forecasting models for city traffic or distance-based tolls (cf. Gullberg, 2015; Transport Analysis, 2019). These enhanced opportunities enable the state to consolidate its position as a central player in information flows.

Furthermore, we find it likely that the state's capacity to govern through *unilateral state communication* also has good prospects to increase in the scenario. The cost of communicating messages to a larger group of individuals is reduced through digital solutions, but on the other hand it may be more difficult for state information to reach out through an increasing information noise (Hood & Margetts, 2007, p. 41ff). Nevertheless, in contrast to the previous scenario, we find it likely that the citizens' acceptance of the state's increased mandate will generate an increased responsiveness to information disseminated by state actors.

All in all, we estimate that the state's nodality-based governing capacity will increase in the scenario Sharing Economy. The assessment is primarily based on the prospect of a technological development combined with a willingness to share information and an increased state mandate within the transport system.

Authority Increases but Fees Become Less Important. In this scenario, the governing capacity for *taxes and fees* in the transport system will likely change. With relatively few privately owned cars and a reduction in the number of kilometres driven, the governing capacity for taxes and fees related to buying, owning and using a car will be reduced. As in the Individualism scenario, the governing capacity of the current fuel tax will be reduced if electric cars make up 70% of the vehicle fleet (cf. Lindberg & Fridstrøm, 2015).

Regarding *regulations for traffic, drivers and vehicles*, the same conditions apply as in the Individualism scenario: the expected digitalisation makes it easier for the state to ensure compliance with several regulations increasing its governing capacity. Other regulations, such as regulations regarding driving licenses, would probably decrease as a result of significantly fewer drivers. While the capacity for regulations concerning taxi legislation may increase if taxi driving is included as part of the widely used MaaS solutions.

Contrary to the scenario above, the urbanisation has continued in this scenario. Although the car is still a necessity for people in rural areas, the scenario is based on aggregation processes. Regardless of such tendencies, we still find it likely that the governing capacity of *regulations for land use and planning* will be comparable with today. When it comes to *data management regulations*, the same reasoning applies as in the Individualism scenario: In a scenario with self-driving cars on levels 3 and 4 and almost entirely self-driving buses, there will be significantly larger amounts of traffic data than is the case today (Bjelfvenstam, 2018, p. 16). Combined with a prevailing positive attitude towards the sharing of data, we find it likely that the governing capacity for data management rules significantly will increase in this scenario.

Summing up, we estimate that the governing capacity for the state's authority increases in this scenario, due to enhanced data access, a general positive attitude towards sharing and increased possibilities to ensure compliance with regulations.

Increased Demand for Road Capacity Strengthens Treasure-based Governing Capacity. In a future with increased concentration of traffic to urban environments, it is likely that the governing capacity of *infrastructure investments* will be somewhat lower than in the previous scenario, but overall, we find it comparable with today. The conditions to govern through infrastructure investment will likely change as a larger part of the investments will be needed in urban infrastructures controlled by local authorities, which the state neither owns nor manages. This could mean that the state increasingly takes on the role of a financier, with consequently increased elements of negotiations with local authorities on terms of state funding.

The scenario involves a rapid development of new services and extensive changes in the behaviour of both people and companies. At a dynamic stage of new solutions, the governing capacity of *R&D efforts* would presumably increase to some extent, although the state's ability to influence through such initiatives is limited in relation to global development and major international players.

Due to the strong urbanisation assumed in the scenario, the demands on the state to equalise and compensate will be high in those parts of the country that are stagnating. The state can use *subsidies* to mitigate the effects of such segregation processes. It is also possible that the difference between shared mobility services, taxi and public transport may become less clear in a future with a prevalent shared economy (Transport Analysis, 2016). It may therefore be justified to implement other forms of subsidies and to other transport modes than is available today (Docherty et al., 2018). Despite such changes, we estimate that the scenario will not mean any change in capacity for this type of governing.

In summary, we assume that the governing capacity for investments in infrastructure and subsidies will be comparable with today, while the governing capacity for R&D efforts will slightly increase. Overall, we find it likely that treasure will continue to be important, with a somewhat increased governing capacity in the Shared Economy scenario.

A Proactive and Decentralised Organisation with Increased Capacity. This scenario suggests a stronger state mandate in the transport system, which we expect will increase the governing capacity of *the scope and structure of the state*. In contrast to the previous scenario, this scenario presents a widespread geographical state that proactively facilitates the achievement of transport policy objectives. We assume that such a development will consequently result in a decentralisation and expansion of the state's operations generating an increased governing capacity.

In contrast to the Individualism scenario, we assume that the governing capacity of *employees and their competencies* will increase in this scenario. This is due to an increased state mandate which will likely result in an increased number of state employees with possibly altered competencies, for example, knowledge of accessibility and shared services will probably be sough-after. Not least to ensure that groups in need of certain care are not regularly neglected. Also, proactive leaders capable of articulating visions of the future play an important role for ensuring transformative capacity (Davis, 2018). It is possible that the increased urbanisation might result in a somewhat decreasing capacity of state infrastructure in relation to infrastructures owned and managed by local authorities. Nevertheless, we find it likely that the overall governing capacity based on the state's *existing infrastructure* will be comparable with today.

All in all, we expect that the governing capacity of the resource organisation will increase in the Sharing Economy scenario. This is primarily based on the assumption of a strengthened state mandate in the transport system, which we estimate will reinforce the entire state organisation.

Conclusions

In this chapter, we have analysed how the governing capacity of currently used policy instruments in the transport sector will be affected in futures of smart mobility. We have drawn upon the four governmental resources of NATO to explore how they will be affected in smart mobility futures. The smart mobility futures were developed through and conceptualised in two contrasting scenarios: *Individualism* and *Shared Economy*. When combining these in the analysis, we condensed the governing capacity of each of these resources and developed an understanding of how they would be affected in each of the scenarios. In Table 2, we present a summarised version of our analysis.

However, it should be noted that the overall summary represented by each of the arrows in the table above is not exhaustive. There are cases where the strength of the governing capacity of a particular instrument is increasing, reduced or remain as it is today. Furthermore, it is important to acknowledge that the analysis does not include reflections on the potential for re-calibration of currently used policy instruments. Due to this omission, the analysis should be regarded as preliminary and as a proposal for further research and discussions, rather than as definitive results.

Looking into each of the NATO resources, we will suggest a number of overall conclusions. *Nodality*, that is, the state's ability to govern by collecting and distributing information, will presumably be challenged in a future of smart mobility.

Table 2. Summary of Combined Governing Capacity of NATO Resources Within Each Scenario.

	Nodality	**Authority**	**Treasure**	**Organisation**
Individualism	↘	↗	↗	↘
Shared Economy	↗	↗	↗	↗

Technology will likely evolve, but the state's capacity to use it in governance is determined by the extent to which the state has access and opportunity to use the information. This is partly determined by the extent to which individuals and organisations are willing to share data, and partly by who owns and controls the data and to what extent the state can exert influence over these actors. Overall, the analysis shows that policy instruments related to *nodality* seem to decrease in the scenario of *Individualism*, while they increase in the scenario of *Sharing Economy*.

Taxes and regulations, which are categorised under *authority*, are among the state's most powerful policy instruments. In fact, in the smart mobility futures, there seems to be completely new opportunities for the state to ensure compliance with traffic rules, and in both scenarios the governing capacity of policy instruments related to the category *authority* generally increases. However, these possibilities depend on whether the state has access to data, for example, on where and how vehicles are driven. In both scenarios, assumptions about traffic development will have a major impact on the governing capacity of taxes and fees. Fuel taxes become less important due to electrification, which means that alternative, distance-based road taxes may become relevant. However, for reasons of integrity, these could potentially be difficult to implement in the *Individualism* scenario.

The governing capacity of policy instruments that make use of the state's *treasure* will also increase in both scenarios. State infrastructure investments will likely continue to be important. One conclusion of the analysis is that governing capacity of investments in R&D will increase in both scenarios due to the potential in new technology and changed behaviour, especially in combination with other instruments such as social networks.

Organisation includes several different aspects, ranging from the government's scope and structure, and employee competences to physical assets in the form of state-owned transport infrastructure. We expect the general governing capacity of *organisation* to be reduced in the scenario of *Individualism* and increased in the scenario of *Shared Economy*. The differing governing capacity of organisation across the scenarios is a result of scenario features that attach different roles to the state's commitment and involvement in society. The role of the state will also be influenced by overall societal and international trends, but it is not apparent from the scenarios what such changes could look like, and it is very difficult to predict.

Furthermore, it should be noted that the resources and policy instruments are interlinked and interdependent. Taxes that we include within *authority* also contribute to *treasure* and *organisation*, while *organisation* is a prerequisite for collecting taxes. Some degree of state *nodality*, and thus access to and control over data, is involved in all other resources, as well as the extent to which other actors compete to be at the centre of informational flows. If the state does not have access and control, it will in some cases be difficult to exercise *authority, treasure and organisation*. For example, increased compliance with traffic rules (*authority*) is dependent on state access to data about vehicle movements (*nodality*). State authorities' access to and control over data is of immense importance, and it is probably essential in order to secure overall governing capacity.

Facing two scenarios where resources and policy instruments may either increase or decrease, the state authorities may choose between two strategies. Drawing upon Offes (2015) notion about governability and ungovernabilty, the state can choose either to focus on keeping and possibly strengthening the governing capacity of those resources that are expected to decrease, or choose to focus on resources and related policy instruments that seem to be robust across different scenarios. This strategy is connected to Lyons' (2018) proposal to choose 'a course of action which has reasonable alignment across multiple futures' (p. 1). While these strategies are principally different, and state authorities might only be able to choose one or the other, the interdependence between the four NATO resources must be acknowledged. Securing the governing capacity of the policy instruments in the categories of *authority* and *treasure* may be insufficient, not least since those instruments are dependent on resources found in the categories of *nodality* and *organisation* too.

All in all, while we have focussed in this chapter on the governing capacity of current policy instruments in two futures of smart mobility, other publications on smart mobility and policy instruments have focussed on the need for specific policy instruments to prevent negative consequences of particular future scenarios (e.g. Swedish Transport Administration, 2018; Transport Analysis, 2017; 2019, see also Pernestål, Engholm, Kristoffersson, & Hammes, 2020, Chapter 3 of this volume). Alongside the general discussion of future governing capacity, the need for specific policy instruments to govern the transition is an important future topic to consider. The demand for instruments is closely linked to the political will to govern the future and on the population's willingness to be governed. Although we suggest that the governing capacity of many policy instruments will either increase or remain the same as today, this does not give much of an indication about the political willingness and thus the possibility of public authorities to apply those instruments.

Acknowledgements

We would like to thank Sweden's innovation agency, Vinnova for funding the research behind this chapter. We also want to thank the other authors of this book for fruitful comments on a previous version of this chapter. The chapter is adapted from a report written in Swedish (Wallsten et al., 2019), that greatly benefitted from comments by Anna Wildt-Persson (The Swedish Transport Administration) and Anna Pernestål (KTH Royal Institute of Technology, Sweden).

References

Bjelfvenstam, J. (2018). *Vägen till självkörande fordon – Introduktion. Slutbetänkande av Utredningen om självkörande fordon på väg [The road to self-driving vehicles - Introduction. Final report of The inquiry about self-driving vehicles on the road]* (SOU 2018:16). Stockholm: Government Office of Sweden.

Cars, G., Healy, P., Madanipour, A., & De Magalaes, C. (Eds.). (2002). *Urban governance, institutional capacity and social milieux*. Oxon, Abingdon: Routledge.

Davis, D. E. (2018). Governmental capacity and the smart mobility transition. In G. Marsden & L. Reardon (Eds.), *Governance of the smart mobility transition* (pp. 105–122). Bingley: Emerald.

Docherty, I., Marsden, G., & Anable, J. (2018). The governance of smart mobility. *Transportation Research Part A: Policy and Practice*, *115*, 114–125.

Field, B. N. (2016). *Why minority governments work*. Houndmill: Palgrave Macmillan.

Finger, M., & Audouin, M. (Eds.). (2018). *The governance of smart transportation systems. Towards new organizational structures for the development of shared, automated, electric and integrated mobility*. Cham: Springer.

Gullberg, A. (2015). *An integrated information and payment platform for urban transport*. (Report from KTH centre for sustainable communications). Stockholm: KTH.

Hood, C., & Margetts, H. (2007). *The tools of government in the digital age*. Basingstoke: Palgrave Macmillan.

Hopkins, D., & Schwanen, T. (2018). Governing the race to automation. In G. Marsden & L. Reardon (Eds.), *Governance of the smart mobility transition* (pp. 65–84). Bingley: Emerald.

Kristoffersson, I., Pernestål Brenden, A., & Mattsson, L. (2017). *Framtidsscenarier för självkörande fordon på väg: Samhällseffekter 2030 med utblick mot 2050 [Future scenarios for self-driving vehicles on the road – Societal effects in 2030 with outlook to 2050]* (VTI notat 18-2017). Linköping: VTI.

Lindberg, G., & Fridstrøm, L. (2015). *Policy strategies for vehicle electrification*. (ITF/OECD Discussion Paper 2015:16). Paris: ITF/OECD.

Lyons, G. (2018). *Handling uncertainty in transport planning and decision making*. (Report of a Roundtable Discussion, London 20 July 2018, Project Report). Bristol: UWE.

Marsden, G., & Reardon, L. (Eds.). (2018). *Governance of the smart mobility transition*. Bingley: Emerald.

Offe, C. (2015). Ungovernability. In A. Azmanova & M. Mihai (Eds.). *Reclaiming democracy: Judgment, responsibility and the right to politics* (pp. 77–86). Abingdon: Routledge.

Painter, M., & Pierre, J. (Eds.). (2005). *Challenges to state policy capacity*. Houndmills: Palgrave Macmillan.

Pernestål, A., & Almlöf, E. (2019). *Scenarier för ny mobilitet och samhällsplanering* [*Scenarios for new mobility and societal planning*] (K2 outreach 2019:7). Lund: K2.

Pernestål, A., Engholm, A., Kristoffersson, I., & Hammes, J. J. (2020). The impacts of automated vehicles on the transport system and how to create policies that target sustainable development goals. In A. Paulsson & C. H. Sørensen (Eds.), *Shaping smart mobility futures – Governance and policy instruments in times of sustainability transitions* (pp. 37–53). Bingley: Emerald.

Stead, D., & Vaddadi, B. (2019). Automated vehicles and how they may affect urban form: A review of recent scenario studies. *Cities*, 92, 125–133.

Swedish Transport Administration. (2017). *Digitaliseringens möjligheter: PM till Nationell plan för transportsystemet 2018–2029* [*The possibilities of digitalization: Memo to National plan for the transport system 2018-2029*] (2017/32405). Borlänge: Swedish Transport Administration.

Swedish Transport Administration. (2018). *Trender i transportsystemet: Trafikverkets omvärldsanalys 2018* [*Trends in the transport system: The Swedish Transport Administration's context analysis 2018*] (2018:180). Borlänge: Swedish Transport Administration.

Transport Analysis. (2016). *Nya tjänster för delad mobilitet.* [*New services for shared mobility*] (2016:15). Stockholm: Transport Analysis.

Transport Analysis. (2017). *Självkörande fordon och transportpolitiska mål.* [*Self-driving vehicles and transport policy objectives*] (2017:20). Stockholm: Transport Analysis.

Transport Analysis. (2019). *Uppkopplade, samverkande och automatiserade fordon, farkoster och system - ett kunskapsunderlag* [*Connected, collaborative and automated vehicles, crafts and systems - a knowledge base*] (2019:8). Stockholm: Transport Analysis.

Wallsten, A., Paulsson, A., Hultén, J., Hedegaard Sørensen, C., Pernestål, A., & Almlöf, E. (2019). *Statlig styrförmåga i framtider med smart mobilitet* [*State governing capacity in smart mobility futures*] (K2 Research Report 2019:9). Lund: K2.

Chapter 10

Micromobility – Regulatory Challenges and Opportunities

Nils Fearnley

ABSTRACT

Shared, dockless micromobility is causing concern across the globe. The phenomenon started with shared bikes and e-bikes. More recently, e-scooters (or electric kickbikes), the focus of this chapter, have flooded cities in unprecedented speed and volume – and have caught virtually every city and competent authority off guard. The failure of current regulatory frameworks to address new challenges posed by e-scooters is explored. This chapter first briefly describes major developments of the shared e-scooter market. It then presents rationales for, and to some extent against, e-scooter regulation as well as policy tools available for e-scooter regulation. E-scooters open the door for new and innovative – and potentially efficient – ways to regulate, including geofencing, zoning, mandatory data sharing and mandatory cooperation. Against this backdrop, the chapter discusses regulatory dilemmas, challenges, opportunities and possibilities.

Keywords: Micromobility; e-scooter; regulation; policy measures; shared mobility; dockless

Introduction

Shared, dockless micromobility has flooded cities around the world in unprecedented speed and volume. The phenomenon started with shared bikes and e-bikes, whose move from docked to dockless operation, has enabled considerable growth

Shaping Smart Mobility Futures:
Governance and Policy Instruments in times of Sustainability Transitions, 169–186

doi:10.1108/978-1-83982-650-420201010

(Nextbike, 2019).[1] More recently, e-scooters – or electric kickbikes[2] – have entered the scene and have sparked enthusiasm as well as concerns. Their popularity can be illustrated by the fact that e-scooters contributed to more than a doubling of shared micromobility use in the US United States between 2017 and 2018 (NACTO, 2019a). In Oslo, a city of just under 700,000 inhabitants, over 16,000 e-scooter trips were made each day only three months after they appeared in the streets (Fearnley & Johnsson, 2019). However, the flip side of this popularity is the nuisance they cause. It has become clear that unregulated shared micromobility generates protests and dissatisfaction and, more broadly, external costs, which may overshadow their benefits. Regulation and governance are at the heart of the solution to the many problems encountered. Applied wisely and thoroughly, regulation may help the industry thrive and provide the travelling public with benefits at minimal societal cost.

This chapter focusses on free-floating shared e-scooters, hereafter referred to as e-scooters, which are offered by e-scooter providers. E-scooters can be picked up anywhere within a specified geographical zone and rented, typically at an upfront cost plus a per-minute fee; they can then be parked anywhere within that zone and be picked up and rented by someone else. For users, the transaction as well as the unlocking and locking of the e-scooter happen on their mobile device. E-scooter providers take care of battery charging, vehicle redistribution and maintenance, and in some instances they remove all vehicles from the streets at night. Their business model is enabled by inter alia, GPS technology, smartphone apps, mobile payment solutions and developments and improvements in battery technology.

Since e-scooters are a new phenomenon, literature and scientific evidence is extremely scarce. This chapter, therefore, relies extensively on policy documents, media and other non-scientific sources, which have been identified by means of media and general internet searches, or they are cited in the scarce literature. Prominence has been given to reports and evidence which is considered impartial, like independent pilot evaluations, international governmental organisations or local government reports.

E-scooter provider Bird was the first company to establish shared dockless e-scooter services in Santa Monica in September 2017 (Yakowicz, 2018). The immediate popularity of e-scooters suggests that they hit a nerve and also met a need and demand. High use and rapid market penetration indicate considerable benefits in terms of social welfare and consumer utility.

Bird and other shared e-scooter providers have grown at an unprecedented pace. Within less than three years of existence, their valuation is often measured in the billions of US dollars (Yakowicz, 2018). E-scooter providers enjoy relatively low capital costs[3] and massive venture capital backing. This combination means

[1]Car sharing has also experienced substantial growth in cases, where they have evolved from a fixed parking to a free-floating system (see e.g. Diana, Chicco, & Ceccato, 2019).
[2]These battery-powered stand-up scooters belong to the PeTs (personal e-transporters) family and should not be confused with Vespa-like scooters and mopeds.
[3]As an illustration, in the consumer end market, an average but decent e-scooter would typically cost between USD 250 and USD 800.

that cities can be flooded with e-scooters literally overnight. Over the past three years, this is exactly what has happened in numerous cities around the world.

The large influx of e-scooters has caught virtually every city they have entered off guard. There are several reasons behind the initial failure of local and central authorities to address the effects of this new transport mode and trade on public rights of way. Firstly, e-scooters operate in a regulatory blind spot (Deighton-Smith, 2018), or at least in the grey areas between local and central government regulation, between commercial use and public space and between bicycling and motorised modes of transportation. Secondly, their introduction has been rapid. Globally, e-scooter providers have expanded from zero to hundreds of cities within a couple of years; locally, cities have experienced the emergence of e-scooters overnight or in very short periods of time. Thirdly, and despite related encounters with other gig and sharing economy disruptions, most authorities had not prepared their legislation with sufficient flexibility to tackle this new phenomenon (see Textbox 1 for the Norwegian case). We saw a similar pattern with the rapid entry and growth of ridesourcing services a few years ago (Zarif, Pankratz, & Kelman, 2019). There are also parallels to the emergence of the Segway in 2002, which surprised regulators and prompted regulatory responses (Mancuso, 2019).

However, public authorities are not the only ones to blame for the current situation. As Dickey (2018) notes, 'Bird, Lime and Spin quickly became known for their strategies of begging for forgiveness rather than first asking for permission' – which is quite similar to the approach taken by Uber (Pelzer, Frenken, & Boon, 2019).

Currently, a majority of cities around the world are pushing back, including Paris, whose streets and public spaces were, until recently, nearly choked by 12 competing e-scooter providers with more than 20,000 vehicles in operation in an anarchic unregulated setting. Cities are implementing all kinds of responses, including temporary (or outright) bans, bylaws, licensing schemes and other regulations in order to tackle problems and advance services.

By looking into the phenomenon of shared dockless e-scooters in the context of regulation, this chapter addresses the claimed benefits against the problems they cause. The objective is to scrutinise rationales and formal justifications for – and to some extent, against – e-scooter regulation. The question of whether or not to regulate e-scooters may not be the most critical one. Regulation is needed. However, the right mix of regulation – and choice of policy instruments – depends on how the problem is framed and understood. This chapter therefore elaborates on features of the e-scooter market which prompt public interventions. A next objective is to explore relevant policy instruments for e-scooter regulation, which may differ substantially from traditional instruments to regulate passenger services for two main reasons. One is the fact that they are not traditional passenger services, which require permits and licenses, but are rather commercial vehicle sharing using public space. The second is related to the fact that technology enables new approaches to regulation.

On this background, this chapter is organised as follows. The next section elaborates on justifications for regulation of e-scooters as well as rationales for the opposite strategy of keeping hands off. A further section explores some

Textbox 1. E-scooters and Legislation in Norway.

The Norwegian case is both illustrative and representative.

Following a legal amendment in 2018 (Lovdata, 2018; MOT, 2018), e-scooters and other PeTs that meet a number of requirements – of which the maximum speed obtained by means of the (electric) motor is central – were to be treated like ordinary bicycles. This meant that these vehicles became available for use by people of all ages, without helmets, on bicycle lanes and on ordinary roads, and that e-scooters could be parked wherever bicycles are allowed. In Norway, this also means that e-scooters could be driven in parks, in pedestrian areas and on pavement.

In early 2019, even before the snow had melted from the streets, the first e-scooter provider entered Oslo. By September 2019, seven different operators were offering e-scooters in the city: Tier, VOI, Circ, Zvipp, Ryde, Lime and Libo. Further, following a tendering process, the Greater Oslo Passenger Transport Authority (PTA) Ruter entered into a strategic cooperation with Tier. A study of the two largest providers in Oslo found that, over a three-week period, they had more than 5,000 e-scooters placed in the streets (Fearnley & Johnsson, 2019). The total supply of e-scooters from all seven providers is unknown.

Oslo municipality's response resembles paralysis. It has not been able to manage or regulate the e-scooter market. Its green – socialist city council has claimed that the central government must take action or provide them with the power to regulate, as they regard e-scooters and bicycles as a regulatory blind spot. At the same time, the liberal – right central government has maintained a hands-off approach, claiming that cities and municipalities are the competent authorities with sufficient power. The latter may hold true, as other Norwegian cities, for example, Trondheim and Stavanger, have banned e-scooters while they prepare the legal basis for operations. These cities regard e-scooters as offering commercial services in a public space, which is subject to application and fees – just like food trucks and other economic activity.

of the more innovative and new policy instruments, or regulatory tools, that become available for the regulation of e-scooters. A next section then discusses opportunities and challenges for e-scooter regulation. A final section concludes this chapter.

E-scooter Regulation

Why Regulate

Beyond the traditional economic rationales for regulating a market, the safeguarding of public interests and societal goals may be relevant justification for interventions in the e-scooter market. In this subsection, we elaborate on the following overall areas as grounds for regulatory intervention:

- market failure;
- use of public space; and
- societal goals

Wider objectives for regulation, such as labour rights and tax evasion, are not considered here.

Market failure. While volumes of books and textbooks address market failures and their causes and remedies, we focus here on what appears to be pertinent to e-scooter regulation. This includes externalities, economies of scale and unfair competition.

Externalities are costs or benefits that do not affect those who take part in a transaction or activity. External costs of transport are not borne (and external benefits are not enjoyed) by travellers who make the travel decision. Emissions are a textbook example of an externality. Emissions cause local and global harm which each motorist does not consider – unless this cost is internalised by means of taxation and fuel duties. Accidents are another important external cost of transportation. A part of the accident risk and associated costs are of course considered by those who travel. These costs are internalised. However, there are wider social costs of accidents, including healthcare, lost production due to sick leave and the grief and suffering of friends and relatives, which are external and hence not considered in the travel decision. It is well established that accident costs are largely not internalised in passenger transport markets and, together with emissions and congestion, constitute the predominant sources of total external costs (see e.g. van Essen et al., 2019).

The chief external cost of e-scooters is traffic safety and accidents.

At the time of this writing, no authoritative sources have established robust estimates of e-scooter accident risk. Both exposure data (vehicle kilometre) and accident data are inherently difficult to obtain and inaccurate. With this uncertainty in mind, e-scooter company Bird registered 37.2 injury reports per million miles travelled (Bird, 2019), which corresponds to 23.1 injuries per million kilometres. Figures in the Portland study (PBOT, 2018) suggest an accident rate of about 136 accidents per million kilometres. Röck (2019) makes a rough estimate for Odense, Denmark, which corresponds to 70 e-scooter accidents per million kilometres; about eight times higher than for bicycles. Preliminary e-scooter data and emergency care unit data from Oslo suggest an accident rate well above 30 per million kilometres. For bicycles in Oslo, in comparison, traffic safety risk has been estimated at eight injuries per one million bicycle kilometres, based on visits to Oslo accident and emergency units (Bjørnskau & Ingebrigtsen, 2015). Thus, despite the great uncertainty around e-scooter injury rates, there is little doubt that the accident risk is considerably higher for e-scooters than for bicycles. Sørensen (2019) compared Norway and Denmark and suggested that the rules and regulations that govern vehicle specification as well as their users (e.g. with regards to age, the drink–drive limit, passengers and helmets) and their use (e.g. access to pavement) are closely tied to the safety performance of e-scooters.

A next justification for regulatory intervention in transport markets is economies of scale. Arnott (1996) provided a convincing first-best case for subsidising taxis, which builds on the same principles that Mohring (1972) demonstrated for public transport: Subsidies improve social welfare because increased supply benefits all users. Hence, there are economies of scale on the consumer side. This may in fact be a justification for subsidies. Additionally, it may also be a justification for restricting the number of operators in an area in order to avoid wasteful competition. Fig. 1 hints at another aspect of economies of scale and density in the e-scooter market: The more vehicles in a city, the more they appear to be used. This makes sense, if it is true, since higher density means increased availability and thus a more attractive service. This in turn means that, all else being equal, a larger e-scooter provider is more profitable.

A final source of market failure relates to other modes and services that compete for the same travelling passengers. In most countries and large cities, passenger transport services – for example, local public transport or taxi services – are thoroughly regulated in terms of market access, prices, fees, safety regulations and so on. For these services to compete with e-scooters on equal terms, regulations that level the playing field may prove necessary in order to avoid unfair competition.

Use of public space. Across the Western world, commercial use of public space is regulated in one way or another. Although e-scooters may operate in a regulatory no man's land, as they fall into the categories of both bicycle and commercial service, few examples outside of purely informal sectors (e.g. buskers and pirate taxis) can be found where commercial activity in a public space is not subject to any form of licensing, operating fees or other regulations. In fact, the use of public rights of way is the reason why cities can, and do, manage and regulate

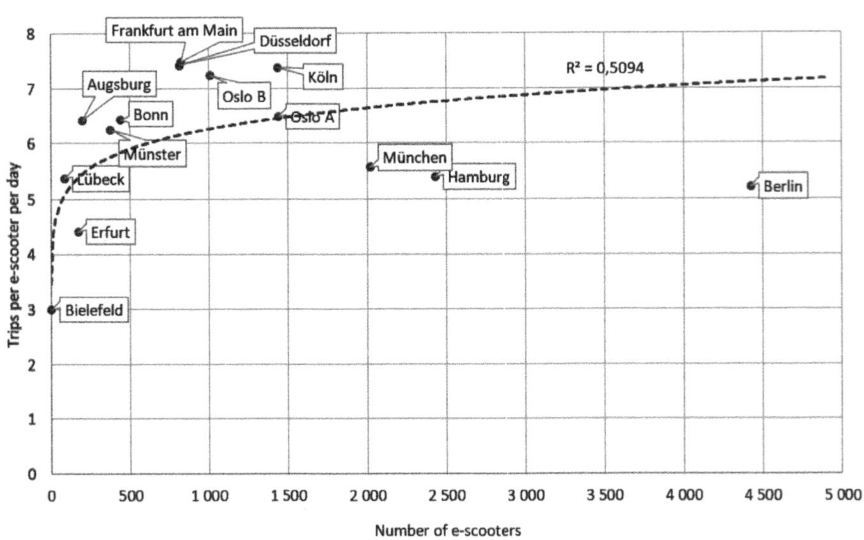

Fig. 1. Trips per E-scooter per Day versus Number of E-scooters in Area.
Sources: *Germany:* Civity (2019) and Oslo, Norway: Fearnley and Johnsson (2019).

e-scooters (NACTO, 2019b). It is thus reasonable and legitimate for cities and their municipalities to take control in shaping their society.

Littering constitutes a major problem wherever e-scooters operate. Furious citizens and representatives of the mobility-impaired complain forcefully about the way e-scooters are parked, as this causes blocking of pathways and pavement and, consequently, increased risk of stumbling and visual pollution. This clearly indicates a need to regulate – and enforce – e-scooter parking.

E-scooter crowding and cluttering is a related problem. Profit-oriented e-scooter providers will, all else being equal, focus their resources on areas with the largest market base and greatest willingness to pay. Hence, e-scooters tend to clump together in city central business districts, at major interchanges and in other central points.

Societal goals. Beyond the need to address market failures, there are wider political and societal goals of relevance for e-scooter regulation. These goals are related to transport as well as to other policy areas. The most urgent transport-related matter of relevance for e-scooter regulation concerns conflicts with other travellers – pedestrians, bicyclists and motorists alike. While traffic rules certainly regulate riding behaviour, regulation must include education of users, clarification of rules, incentivising of adherence to rules and enforcement.

In addition, within the transport sector, policies for sustainable and efficient transport systems emphasise the need to curb car use and reduce congestion. E-scooters can potentially help with this. Table 1 cites some references which have studied the degree to which e-scooter rides replace car trips. Note that several of these references are provided by the industry itself and cannot be regarded as neutral. The picture is fairly stable, at least for North American locations: about one-third of e-scooter trips appear to replace a car trip. It means that e-scooters can potentially help alleviate some of the problems which cars bring with them. More generally, it matters where the e-scooters are located. Regulation can determine that a certain number, or share, of e-scooters be located in car-reliant areas. This way, vehicles can be spread more evenly in the geography and with focus on locations where they most effectively replace car trips.

Along the same lines, e-scooters can divert demand away from public transport, walking and cycling. Indeed, most surveys find that e-scooter trips predominantly replace walking. However, there is also evidence that e-scooters can complement public transport and offer a solution to the fist/last mile problem (ITDP, 2018). A fair share of e-scooter trips are in fact trips to and from public transport. Policy instruments, which make e-scooters a reliable and attractive first/last mile solution, are needed. Again, geographical zoning regulation may prove effective.

Finally, e-scooters generate enormous amounts of supply and demand data, which may prove useful for a number of transport-related policy purposes, including public transport network planning. E-scooter data are nearly costless to produce and share, but access to data may need to be regulated.

Outside the field of transport policy, e-scooters can (positively or negatively) impact and contribute to public health, social wellbeing, equity, economic cohesion and an inclusive society. Regulation may be needed to stimulate benefits or minimise disbenefits. Regarding equity, the evidence is mixed. On the one side,

Table 1. Cited Impact of E-scooters on Car Use and Mode Shift.

Location (Source)	Quoted Impact on Mode Shift
Portland (PBOT, 2018)	34% of Portland riders and 48% of visitors took an e-scooter instead of driving a personal car or using Uber, Lyft, or taxi
San Francisco (SFMTA, 2019)	Up to 40% of scooters trips may be replacing trips that would otherwise be made using private automobiles
Denver (Denver Public Works, 2019)	If a dockless vehicle had not been available, 22% would most likely have taken a taxi, Lyft, or Uber rideshare; and 10% driven a motor vehicle/car (person vehicle, carshare vehicle, rented vehicle and other)
Santa Monica and San Diego (Ajao, 2019)	30% of riders report using e-scooters to replace car rides on their most recent trip
Brussels (Bruzz, 2019)	A negligible 3% would use car without the scooter. Taxis and Ubers account for another 3%
Oslo (Berge, 2019)	5% report that the last e-scooter trip replaced car
St. Louisa[a]	Nearly 40% of St. Louis Lime riders use electric scooters to replace car rides
Minneapolis[a]	38% of trips replaced a personal car, rideshare, or taxi ride according to respondents
Christchurch[a]	1,000,000+ scooter trip are equivalent to an estimated 293,000 car trips avoided
Brussels[a]	25% report using a Lime scooter to replace a trip by car during their last ride

[a]Lime surveys reported on Lime's blog 2nd *street* (https://www.li.me/second-street).

e-scooter users resemble typical early adopters, that is, young, high-income, highly educated men (see Rogers, 1962; an analysis of early adopters of battery electric cars is provided by Figenbaum et al., 2015). At the same time, evidence from the United States suggests that dockless vehicle-sharing in general, and e-scooters in particular, attract usage by and provide an attractive mobility option for people from all walks of life, for affluent and low-income urban areas alike.

… *And Why Maintain a Hands-off Approach*

For the sake of balance, this section emphasises the fact that regulation may not always be the preferred solution. Two related principles, adopted by the European Commission, accentuate this. The first is the principle of subsidiarity – that is, that decisions should be made at the lowest political level necessary. This means, for example, that regulations that are best managed at a municipal level should not be dealt with at national or supranational levels. This is a case for higher level governments to *not* regulate. The second is the principle of proportionality, which requires that regulations should not exceed what is strictly necessary to achieve the objectives.

ITF (2019a) emphasises and reiterates these principles when urging minimum necessary interference and that the minimum regulatory costs or barriers be imposed on e-scooters. A transparent process of regulatory impact assessment (RIA; see OECD, 2012) should carefully make certain that 'any regulatory interventions avoid distorting markets or unnecessarily restricting the development of innovative market offers' (ITF, 2019a, p. 9). In other words, regulatory intervention must be justified. If the goal can be achieved with less intrusive actions or at less cost to e-scooter providers by other means, then the regulation in question should be abandoned. ITF (2019a) further stresses that 'Minimising regulatory barriers is particularly important where new modes and business models with uncertain viability, such as dockless bikeshare and e-scooters are concerned' (p. 7) and recommend, in general, light-handed and supportive regulatory approaches.

As a general note, the fact that the e-scooter industry itself so actively demands 'rules of the game' and regulations would in normal circumstances raise a red flag. Industries may demand regulation in order to protect themselves, for example, from competition, and the result may be a situation of regulatory capture, whereby regulation protects the industry rather than the public good (see Stigler, 1971). In such a case, less regulation improves social welfare.

Policy Instruments and Regulatory Tools

What makes e-scooters an interesting subject for exploring policy instruments is the fact that traditional approaches to passenger transport regulation, for example, entry regulation, price regulation, licensing, fleet caps and prohibition, remain largely ill-advised. E-scooters represent a whole new set of challenges, which require new regulatory approaches and, in part, offer a new set of possibilities for targeted regulation – due, for example, to the fact that they are constantly online and geopositioned. The bulk of this section is therefore devoted to new possibilities and responses to new challenges.

Geofencing is probably the most novel opportunity enabled by e-scooters for the regulation of passenger transport. E-scooter providers themselves already use geofencing extensively to define 'go zones', 'no-go zones' and 'slow zones', limit the geographical area where e-scooters can be parked, and so on. For example, providers may opt to prohibit parking near river banks to minimise the risk of e-scooters being thrown into the water, to prohibit riding in cemeteries or to reduce speeds in pedestrianised areas.

Geofencing in this way is also available for regulators. They may also define fleet caps (or minimum supply) in different geographical zones of a city. In Denver, for example, zoning, fleet restriction and geofencing are combined:

> To incentivize a more equitable distribution of the vehicles, operators wishing to deploy the maximum number of vehicles are required to deploy at least 100 vehicles in 'opportunity areas'

> outside the city core [...] In order to better integrate the dockless vehicles into the exiting transportation system, operators are required to rebalance vehicles to areas near transit stops at the beginning of each day. (Denver Public Works, 2019, p. 1)

Geofencing can also be used in combination with other regulatory measures. For example, geofencing combined with price regulation can contribute to better integration with local public transport and to solving the first/last mile problem. According to Macku (2019), of e-scooter provider Dot, it is entirely possible to reduce fares for trips that start or end at a public transport station. ITDP (2018, 2nd paragraph) follows this line and suggests that 'cities could work with operators to subsidize scooter and bikeshare rides that start or end at transit using common payment options'.

Even more sophisticated opportunities arise when geofencing regulation happens in real time. E-scooter demand, supply and parking may be regulated differently at different times of the day or the week, or an incident may prompt a temporary need to ban e-scooters in an area. Similarly, real-time regulation of e-scooter parking can cater for the changing need for curbside space in downtown areas between, say, goods deliveries during daytime, commuting during peak hours and taxi services on a Friday night. This can even be automated, that is, regulation by algorithm (ITF, 2019b).

Unfortunately, the GPS systems currently used can be inaccurate up to 100 metres in dense cities with tall buildings. For the time being, fine-grained regulation of, for example, speeds on pavement versus on the road alongside, or parking on different pavement sections, must be enforced by police or traffic wardens.

The most intrusive regulatory response to e-scooters is of clearly to *ban* them. No doubt, this is the preferred solution for many citizens. General prohibition is, however, not particularly productive, nor does it reflect the considerable welfare benefit and utility enjoyed by a large number of happy users. Still, a large number of cities have banned e-scooters temporarily while they figure out and design the rules of the game.

Less intrusive than bans, but still a considerable burden for a for-profit operator, is the *capping of fleet sizes* – which has become a widespread practice. A fleet cap may appear a reasonable way to limit the total number of vehicles or the number of vehicles each e-scooter provider may offer. A cap may be conditional and include rules for when fleets can be expanded or must be reduced. Examples of such rules include the possibility of expanding the fleet when usage per vehicle per day exceeds a predefined number, or to reduce the fleet if average usage falls below a certain threshold. Regarding fleet caps, it is worth noting that ITF (2019a) considers caps particularly negative, and Zarif et al. (2019) argue that fixed caps are inferior to outcome-based rules. This is also supported by the evidence presented in Fig. 1, which indicate economies of scale and density.

Various forms of market *entry regulations* can be used to control the number of competing e-scooter operators. This can take the form of ordinary competitive tendering processes where operators bid for one or more contracts. When evaluating bids for concessions, however, ITF (2019a, p. 39) notes that there exist no established decision criteria. An alternative is licence auctioning. For both of these procurement methods, an appropriate contract duration must be set. The economic life of an e-scooter is currently assumed to be well under 6 months, suggesting that contract durations should be short. However, the industry tends to prefer permits over procurement – an example of which is the 2019–2020 pilot in Portland, which organised an application process for permits to operate (PBOT, 2019).

Charges and fees serve a number of purposes. Given that other commercial activity on public rights of way is subject to charges and fees (sometimes including application fees), the imposition of the same fees on e-scooter providers only levels out the playing field. Fees also contribute to cover the cost of regulation and enforcement. However, it is more interesting to consider charges and fees as incentives for e-scooter providers to behave in ways that align with policy aims, in which e-scooter companies trade off fees against benefits and revenues when choosing their operation parameters (e.g. number of vehicles, location and pricing). In this way, pricing strategies can have a dynamic effect. Consider as a simple example the different incentives provided by a flat annual fee of USD 500 per vehicle versus a daily fee of USD 2 per vehicle. The former provides little incentive to remove e-scooters on rainy or snowy days, while the latter provides an incentive to remove damaged scooters and excessive supply. However, the former incentivises vehicle longevity (i.e. scooters that last at least one year).

Herrman (2019, p. 46) showed that permit fees are most frequently used in United States cities, followed by daily fees per scooter, application fees and renewal fees. An example is, again, the 2019 Portland pilot fee structure, which consists of a one-off permit application fee of USD 500, a per-scooter pilot permit fee of USD 80, a street use surcharge of USD 0.25 per trip and a right-of-way use surcharge of between USD 0.05 and 0.20 per trip, depending on the area (PBOT, 2019). There are clearly some built-in incentives that may affect an e-scooter provider's operation parameters.

Penalties are related to fees and charges. City authorities may apply penalties in numerous circumstances, including improper parking, failure to remove vehicles within a predefined reaction period, failure to educate users and missing safety equipment. Monitoring and enforcements are prerequisites for an efficient penalty scheme.

Compulsory data sharing is one of the big success stories of e-scooter regulation. E-scooters generate enormous amounts of use and supply data with unprecedented quality, completeness and detail. These data give new insights into urban travel patterns and city use – with benefits far beyond the e-scooter industry. In fact, e-scooter data represent one of the factors that provides this new service much-needed legitimacy and support. To date, the evidence suggests that e-scooter providers do not share detailed operational data with cities and the public voluntarily. Compulsory data sharing therefore can and should be an essential part of all operating licences. A user survey requirement may be an integral part of this (Yanocha, 2018). LADOT (2019) has developed and designed a Mobility Data Specification (MDS), which sets out the requirements for data sharing. MDS is available on GitHub (GitHub, 2019) and is currently used by more than 50 United States cities and dozens of other cities (Open Mobility Foundation, 2019).

A successful regulatory strategy, *pilot schemes,* concludes this section. Successful pilot programmes and trial schemes in the United States have demonstrated the benefits of *regulatory sandboxes*, where new e-scooter policies can be tested for up to a year at a time. Learning and adapting are at the heart of these pilot schemes, as operators and authorities can cooperate and test various solutions and evaluate their outcomes. With the pace of change in micromobility in general and considering the limited life expectancy of a micromobility vehicle, pilots of limited duration may in fact be the preferred status quo.

Discussion

Despite considerable protest and complaint, e-scooters are extremely popular – especially among the younger generations. Populus (2018, p. 3) notes that, 'Based on data from over 7,000 people in major U.S. cities, a majority of people (70%) view electric scooters positively'. The explosive growth and their omnipresence wherever they enter a market suggest a success story for a new mobility service that provides significant benefits to its users. Adding to that, there is some evidence to suggest that e-scooters can contribute to a more equal and inclusive society. Among women, they are relatively more popular than city bike schemes, and e-scooters enjoy relatively higher adoption rates by lower-income groups (Populus, 2018). They also offer an accessible transport alternative in mobility-deprived areas.

Nevertheless, a laissez-faire approach to e-scooters has proven unsustainable. There appear to be two key reasons why size is important in this context and drives operators to flood the market with vehicles. One is economies of scale, explained above. The dynamics of scale economies create a situation in which e-scooter providers struggle and compete to become larger and thus the preferred provider. The other is the business model of e-scooter start-ups and operators, outlined in ITF (2019a). It holds that venture capital backed gig economy start-ups like most e-scooter operators do not follow a traditional business model of seeking (short-term) profits from operations. Rather, their objective is to expand their market penetration and increase their market shares in order to inflate stock value (ITF, 2019a, p. 40).[4]

The dilemma then arises: On the one hand, a large number of e-scooter vehicles cause concerns related to littering, cluttering and obstruction – clearly an untenable situation. On the other hand, a large number of vehicles are also central to user benefit and the attractiveness of the services, in addition to the commercial benefits of large and high density of supply. Nonetheless, supply must be managed, in one way or another. Even e-scooter operators are pro-regulation (Macku, 2019) and ask for rules of the game. The previous section outlined some available regulatory possibilities.

---

[4]Despite this observation, however, e-scooter operators VOI and Tier have announced that they made a profit in Oslo (Wasberg, 2019) about half a year after their services commenced.

However, the fact that e-scooters are an infant and largely unprofitable industry in an extremely dynamic and innovative market suggests that the amount and intensity of regulations should be kept at a minimum. ITF (2019a, p. 7) recommends that:

> Regulation should reflect an essentially permissive and facilitative approach to innovation, which accepts market disruption, rather than seeking artificially to slow or impede the adoption of new business models and technologies. This does not imply inaction where there is a clear need to protect consumers from the risk of significant harm. Minimising regulatory barriers is particularly important where new modes and business models with uncertain viability, such as dockless bikeshare and e-scooters are concerned.

Deighton-Smith (2018) also warns about the risk that regulation will choke a vulnerable industry and that it may distort its development. Hence, regulations should seek to minimise market distortions, and not inhibit innovation. In contrast to this, Pankratz, Nuttall, Eggers, and Turley (2018) hold that 'regulation versus innovation' can be a false contrast. Well-formulated regulation, they argue, could in fact act as a catalyst for innovation and development.

Further, there is vast variation in the regulatory responses taken by cities around the globe during the first two years of the commercially shared e-scooter's existence. We have witnessed the whole spectrum, from outright bans to a laissez-faire approach. More worrisome, perhaps, are the shifting and ad hoc policy responses that echo shifts in public opinion. In Norway, as an illustration, e-scooters have been deemed legal and hence unregulated in Oslo, while they have been banned in other cities. For e-scooter operators in Norway, this lack of clarity represents considerable regulatory risk. Clearly formulated regulations with clearly communicated plans for revisions and amendments will help stabilise matters and give e-scooter providers some predictability. (That said, however, the e-scooter companies themselves represent considerable disruption and instability.)

While authorities should seek to reduce regulatory risk, this early stage of innovation and disruption clearly calls for a flexible approach to regulation. Yesterday's problems are addressed with tomorrow's technology and practices. What may seem a passable regulatory response to a problem may produce unexpected and adverse side effects (see Justen, Fearnley, Givoni, & Macmillen, 2014). Services evolve and change rapidly, and swift and flexible regulatory responses are needed. Therefore, a conclusive set of rules and regulations is neither possible nor a goal. Instead, all this emphasises the need for a built-in flexibility of regulatory practices and, equally important, the need to plan evaluations and amendments. Regulatory sandboxes, trials and pilots are excellent strategies for this, as they build partnership of mutual trust and a mutual understanding of objectives and strategies – and they encourage innovation. Innovation can indeed also take place on the regulator side.

In addition to restrictive regulations, authorities may do well to implement positive and enabling policies (ITF, 2019a). One such enabling action would be the provision of cycle paths. All evidence shows that e-scooter riders prefer to use bike lanes so as to avoid conflicts with pedestrians and motorists. More bike lanes will contribute to many cities' wider transport policy objectives regarding increased active transport and increased traffic safety. Another enabling policy would be to clarify parking rules and provide guidance – and space – for e-scooter parking.

The industry can and does team up and agree on industry code of conduct. However, the regulatory initiative and leadership must be taken by competent authorities – which, in the case of e-scooters, would largely be municipal city authorities. In the U.S., cities currently join forces and organise themselves in different ways. The formation of the Open Mobility Foundation (Hawkins, 2019) is one prominent example of how coalitions can help authorities build competence and exchange knowledge and practices. Better, and more efficient, regulation will likely result.

Conclusions

Cities have been caught flat-footed and off guard by the sudden influx of e-scooters over the past two years. Initially operating in a regulatory blind spot, the unleashed e-scooter market has clearly demonstrated a need for regulation.

The following points stand out as takeaway messages:

- There is, in total, a strong case for the regulation of e-scooter markets. The need for regulation stems from market failures, the use of public rights of way and the protection of wider societal goals. However, regulators should take only the minimum actions necessary to achieve the objectives, and be facilitative.
- The fact that e-scooters are online and geopositioned opens up new opportunities for regulation. These include geofencing and a range of related opportunities, as well as data sharing requirements that will enable new and deeper understanding of travel patterns.
- Pilot schemes and regulatory sandboxes are well-suited approaches to testing and evaluating regulatory response actions.
- Rapidly emerging and changing technology requires a flexible approach to regulation, with built-in plans for revisions and amendments.

Despite initial, and considerable, negative public attention, there are reasons to be optimistic and to envision a role for e-scooters as part of the solution to urban mobility. Shared e-scooters only entered the scene in late 2017. Two years on, at the time of this writing, e-scooter companies have already rolled out versions two, three, or four, of their vehicles. A number of concerns about e-scooters have already been – or are on their way to being – addressed, including longevity of the vehicles and their components, and certain comfort and safety issues.

The mere fact that e-scooters represent something new, as opposed to, for example, cars, opens a window of opportunity for early and restrictive regulatory responses, which may not be possible or acceptable for other modes. For example, geofencing of car speeds and car parking is technically possible. But cars have been around and enjoyed their freedoms for so long that many policymakers would hesitate to impose similarly restrictive policies for cars as they consider for e-scooters. Restrictive car policies are predictably accompanied by protests, while e-scooter restrictions are in general welcomed.

There are also reasons to believe that public acceptance, e-scooter safety, parking, riding behaviour and littering will improve. The key to this involves adaptation, learning and time. In San Francisco, SFMTA (2019, p. 2) reports that 'complaints about sidewalk riding and improper parking were significantly reduced under the Pilot'. In the 2018 Portland pilot, accident rates increased initially but were back to pre-pilot levels at the end of the pilot period (PBOT, 2018). Considerable learning and adaptation around the e-scooters is taking place, as well. Although they may not wholeheartedly welcome them, motorists and pedestrians are slowly getting used to the presence of e-scooters. And, importantly, city authorities are, one by one, entering the scene and, increasingly, taking a leadership role. Public opinion may also improve, slowly. Today, a parked passenger car occupies roughly 20 times the space of an e-scooter and they kill thousands, yet they attract far less negative attention. With time, this attitude may well change in favour of e-scooters.

Acknowledgements

The research leading to this text was financed by the Research Council of Norway via the REGSMART project (project number: 283327); funding was also provided by an e-scooter knowledge-building project sponsored by Ruter, Bymiljøetaten, Helsedirektoratet and Statens vegvesen vegdirektoratet, from which one chapter in a research report (in Norwegian) will share some of the contents of this chapter. The author is grateful to Jørgen Aarhaug, Siri Hegna Berge and Espen Johnsson for useful discussions and helpful suggestions.

References

Ajao, A. (2019). Electric scooters and micro-mobility: Here's everything you need to know. *Forbes*. Retrieved from https://www.forbes.com/sites/adeyemiajao/2019/02/01/everything-you-want-to-know-about-scooters-and-micro-mobility/#49349d945de6. Accessed on November 21, 2019.

Arnott, R. (1996). Taxi travel should be subsidized. *Journal of Urban Economics*, *40*, 316–333.

Berge, S. H. (2019). *Kickstarting micromobility – A pilot study on e-scooters*. TOI-report 1721/20. Retrieved from https://www.toi.no/getfile.php?mmfileid=50932. Accessed on November 25, 2019.

Bird. (2019). A look at e-scooter safety. Examining risks, reviewing responsibilities, and prioritizing prevention. Retrieved from https://www.bird.co/wp-content/uploads/2019/04/Bird-Safety-Report-April-2019-3.pdf. Accessed on September 11, 2019.

Bjørnskau, T., & Ingebrigtsen, R. (2015). *Alternative understandings of risk and exposure*. TOI report 1449/2015. Retrieved from https://www.toi.no/getfile.php?mmfileid=41817

Bruzz. (2019). Elektrische step vervangt vooral voetgangers en amper auto's [Electric scooter mainly replaces pedestrians and hardly cars]. Retrieved from https://www.bruzz.be/analyse/elektrische-step-vervangt-vooral-voetgangers-en-amper-autos-2019-06-19. Accessed on November 21, 2019.

Civity. (2019). *E-Scooter in Deutschland: Ein datenbasierter Debattenbeitrag*. Civity Management Consultants GmbH & Co. KG. Retrieved from http://scooters.civity.de/. Accessed on September 10, 2019.

Deighton-Smith, R. (2018). *The economics of regulating ride-hailing and dockless bike share*. Discussion Paper, Roundtable 175, International Transport Forum, Paris.

Denver Public Works. (2019). Denver dockless mobility program pilot interim report – February 2019, Retrieved from https://www.denvergov.org/content/dam/denvergov/Portals/705/documents/permits/Denver-dockless-mobility-pilot-update-Feb2019.pdf. Accessed on September 15, 2019.

Diana, M., Chicco, A., & Ceccato, R. (2019). Car-sharing: What can we learn from current business models and which model to choose for your city? Presentation to POLIS Governance & Integration Working Group meeting, 26 February, Brussels, Belgium.

Dickey, M. R. (2018). The electric scooter wars of 2018. *Techcrunch*. Retrieved from https://techcrunch.com/2018/12/23/the-electric-scooter-wars-of-2018/. Accessed on September 15, 2019.

van Essen, H., van Wijngaarden, L., Schroten, A., Sutter, D., Bieler, C., Maffi, B. M., ... El Beyrouty, K. (2019, January). *Handbook on the external costs of transport Version 2019*. Luxembourg: Publications Office of the European Union.

Fearnley, N., & Johnsson, E. (2019). Elsparkesyklene – nye fakta om tilbud og bruk i Oslo. *Samferdsel*, September 17. Retrieved from https://samferdsel.toi.no/hjem/article34312-98.html

Figenbaum, E., Fearnley, N., Pfaffenbichler, P., Hjorthol, R., Kolbenstvedt, M., Jellinek, R., ... Iversen, L. M. (2015). Increasing the competitiveness of e-vehicles in Europe. *European Transport Research Review*, *7*, 28. doi:10.1007/s12544-015-0177-1

GitHub. (2019). *CityOfLosAngeles/mobility-data-specification*. Retrieved from https://github.com/CityOfLosAngeles/mobility-data-specification. Accessed on September 18, 2019.

Hawkins, A. J. (2019). US cities are joining forces to figure out what the hell to do with all these scooters. *The Verge*. Retrieved from https://www.theverge.com/2019/6/25/18715977. Accessed on September 15, 2019.

Herrman, M. (2019). *A comprehensive guide to electric scooter regulation practices*. A report submitted in partial fulfillment of the requirements for the degree Master of regional and community planning. Department of Landscape Architecture and Regional & Community Planning College of Architecture, Planning and Design, Kansas State University.

ITDP. (2018). *E-scooters could be a last mile solution for everyone*. Institute for Transportation & Development Policy. Retrieved from https://www.itdp.org/2018/12/14/e-scooters-last-mile-solution/. Accessed on September 10, 2019.

ITF. (2019a). *Regulating app-based mobility services. Summary and conclusions*. Roundtable 175, International Transport Forum/OECD.

ITF. (2019b). *Governing transport in the algorithmic age*. International Transport Forum/OECD Corporate Partnership Board Report.

Justen, A., Fearnley, N., Givoni, M., & Macmillen, J. (2014). A process for designing policy packaging: ideals and realities. *Transportation Research Part A: Policy and Practice*, *60*, 9–18.

LADOT. (2019). Dockless mobility program. Retrieved from https://ladot.io/programs/dockless/. Accessed on September 18, 2019.

Lovdata. (2018). *Forskrift om endring i forskrift om krav til sykkel* [*Regulation on amendment to regulation on bicycle requirements*], *FOR-2018-04-09-545*.

Macku, M. (2019). *Flash. Connect the dots*. Presentation to POLIS Governance & Integration Working Group meeting 26 February, Brussels, Belgium.

Mancuso, K. (2019, January). The rise of electric scooter regulations. *The Regulatory Review*. Retrieved from https://www.theregreview.org/2019/01/03/mancuso-electric-scooter-regulations/. Accessed on September 15, 2019.

Mohring, H. (1972). Optimization and scale economies in urban bus transportation, *American Economic Review*, *62*, 591–604.

MOT. (2018). *Små elektriske kjøretøy blir likestilt med sykkel* [*Small electric vehicles will be treated in pari passu with bicycles*]. Press release, Ministry of Transport and Communications, 10 April 2018.

NACTO. (2019a). *84 million trips in 2018: Shared micromobility in the U.S.: 2018*. National Association of City Transportation Officials and Better bike share partnership. Retrieved from https://nacto.org/wp-content/uploads/2019/04/NACTO_Shared-Micromobility-in-2018_Web.pdf. Accessed on September 10, 2019.

NACTO. (2019b, September). *Guidelines for regulating shared micromobility. Version 2*. National Association of City Transportation Officials. Retrieved from https://nacto.org/wp-content/uploads/2019/09/NACTO_Shared_Micromobility_Guidelines_Web.pdf. Accessed on April 11, 2020.

Nextbike. (2019). *Nextbike original bike sharing*. Presentation to POLIS Governance & Integration Working Group meeting 26 February 2019, Brussels, Belgium.

OECD. (2012). *Recommendation of the council on regulatory policy and governance*. Paris: OECD Publishing. https://doi.org/10.1787/9789264209022-en

Open Mobility Foundation. (2019). Retrieved from https://www.openmobilityfoundation.org/. Accessed on September 18, 2019.

Pankratz, D. M., Nuttall, K., Eggers, W. D., & Turley, M. (2018). Regulating the future of mobility. *Deloitte Insights*. Retrieved from https://www2.deloitte.com/us/en/insights/focus/future-of-mobility/regulating-transportation-new-mobility-ecosystem.html. Accessed on September 15, 2019.

PBOT. (2018). *2018 E-Scooter findings report*. Portland Bureau of Transportation. Retrieved from https://www.portlandoregon.gov/transportation/article/709719. Accessed on April 11, 2020.

PBOT. (2019). *Shared electric scooters permit application*. Portland Bureau of Transportation. Retrieved from https://www.portlandoregon.gov/transportation/article/726366

Pelzer, P., Frenken, K., & Boon, W. (2019). Institutional entrepreneurship in the platform economy: How Uber tried (and failed) to change the Dutch taxi law, *Environmental Innovation and Societal Transitions*, *33*, 1–12. https://doi.org/10.1016/j.eist.2019.02.003

Populus. (2018). *The micro-mobility revolution: The introduction and adoption of electric scooters in the United States*. Populus Research Report. Retrieved from https://research.populus.ai/reports/Populus_MicroMobility_2018_Jul.pdf. Accessed on September 15, 2019.

Rogers, E. M. (1962). *Diffusion of innovations*. Glencoe, Illinois: Free Press.

Röck, N. D. (2019). *Pressemeddelelse fra ulykkes analyse gruppen: Skader ved brug af el-løbehjul* [*Press release from accident analysis group: injuries from e-scooter use*]. Odense: Odense.

SFMTA. (2019). *Powered scooter share mid-pilot evaluation*. San Francisco Municipal Transportation Agency. Retrieved from https://www.sfmta.com/sites/default/files/reports-and-documents/2019/04/powered_scooter_share_mid-pilot_evaluation_final.pdf. Accessed on September 15, 2019.

Sørensen, M. W. J. (2019). *Elsparkesykkel trolig farligere i Norge enn i Danmark* [*E-scooters are likely more dangerous in Norway than in Denmark*]. Samferdsel, August 4. Retrieved from https://samferdsel.toi.no/hjem/elsparkesykkel-trolig-farligere-i-norge-enn-i-danmark-article34260-98.html. Accessed on September 15, 2019.

Stigler, G. J. (1971). The theory of economic regulation. *The Bell Journal of Economics and Management Science*, *2*(1), 3–21. doi:10.2307/3003160

Wasberg, E. S. (2019). *Millionmilepæl for Voi: Nå hevder de og konkurrenten Tier at selskapene er lønnsomme i Oslo* [*Milestone for Voi: now they and competitor Tier claim that they make profits in Oslo*], Dagens Næringsliv. Retrieved from https://e24.no/boers-og-finans/i/RRaLe8/millionmilepael-for-voi-naa-hevder-de-og-konkurrenten-tier-at-selskapene-er-loennsomme-i-oslo. Accessed on August 13, 2019.

Yakowicz, W. (2018). *14 Months, 120 Cities, $2 Billion: There's never been a company like bird. Is the world ready? Inc. Magazine*. Retrieved from https://www.inc.com/magazine/201902/will-yakowicz/bird-electric-scooter-travis-vanderzanden-2018-company-of-the-year.html. Accessed on September 15, 2019.

Yanocha, D. (2018). *Optimising new mobility services*. International Transport Forum Discussion Papers, Roundtable 175, Paris: OECD Publishing,

Zarif, R., Pankratz, D. M., & Kelman, B. (2019). *Small is beautiful: Making micromobility work for citizens, cities, and service providers. Deloitte Insights*. Retrieved from https://www2.deloitte.com/us/en/insights/focus/future-of-mobility/micro-mobility-is-the-future-of-urban-transportation.html. Accessed on September 14, 2019.

Chapter 11

Smart Public Transport in Rural Areas: Prospects, Challenges and Policy Needs

Fredrik Pettersson and Jamil Khan

ABSTRACT

There is a heavy dependence on cars for people living in rural areas and small towns. The countryside has so far been left out of the transition to carbon-free transport, and public transport shares are low in rural areas. New information and communication technology (ICT) solutions and autonomous vehicles (AVs) have the potential to improve the conditions for public transport in rural areas, as they may increase efficiency and reduce costs. Still, these technological novelties are rarely tested in rural settings and policy focus and pilot tests have occurred almost exclusively in cities. The aim of this chapter is to explore the conditions and challenges for public transport in rural areas through ICT and AVs. The authors will discuss how policy focus needs to change to increase attention to rural areas and give suggestions on concrete policy measures that can be used. In the chapter, the authors draw empirically on results from two research projects in Sweden about the conditions for public transport in rural areas and ongoing tests with new ICT solutions.

Keywords: Public transport; autonomous vehicles; ICT; rural transport; transport policy; shared rural mobility

Introduction

People in rural areas and small towns are very dependent on cars for their mobility. The countryside has so far been left out of the transition to carbon-free transport, which has been focussed mainly on cities. While public transport has increased substantially in Sweden in the last ten years, mainly in and between cities, rural areas have witnessed cutbacks, due to high costs, which has led to decreased accessibility and increased dependence on cars.

Shaping Smart Mobility Futures:
Governance and Policy Instruments in times of Sustainability Transitions, 187–201

doi:10.1108/978-1-83982-650-420201011

There are several reasons for emphasising the need for directing policy focus to address the lack of alternatives to the car in rural areas (Winslott Hiselius et al., 2019). One reason is that policy measures aimed at reducing transport sector emissions, such as increased fuel taxes, or the current Swedish bonus malus system for new vehicles, will risk increasing the divide between the urban and the rural. Obviously, any economic policy measure intended to make car use more expensive will have a negative impact in areas where there is no feasible alternative to the car. This is especially the case since the divide between rural and urban areas is clearly illustrated by the differences in income and economic status between urban and rural populations. Using policy 'sticks' to curb the use of cars in contexts where there is no alternative will also, by default, have a limited effect. Additionally, it may be counterproductive for securing political support for climate transition policies. The yellow vests in France, and the more docile, but yet quite vociferous gasoline rebellion in Sweden (demanding lower taxes on petrol and diesel) are examples of how the techno-rational approaches of introducing the most (cost) efficient policies may trigger grassroots movements protesting policies part of the governance towards low carbon transport. In both of the above cases, the protests originated in non-urban areas, but have since become more widespread phenomena channelling a dissatisfaction with what are seen as unjust, regressive policies (in general, but also specifically transport policies) with an unfair distribution where 'urban elites' are favoured over rural residents.

As such, securing popular support for policy measures to limit transport sector climate emissions may involve a different governance approach based on a holistic view of how to facilitate changing travel behaviour in both rural and urban areas. Here, the introduction of new transport services involving information and communication technology (ICT) and autonomous vehicles (AVs) in rural areas could play an important role. However, for this to happen we argue that there is a need for a change in policy focus, and perhaps more importantly a change in the governance approach to address the problems with transport in rural areas.

New ICT and AVs have the potential to improve the conditions for public transport in rural areas since they may increase efficiency and reduce costs, but there are also many challenges and barriers to put these technologies to use. Still, the technological novelties are rarely tested in rural settings and policy focus and pilot tests have occurred almost exclusively in cities. In fact, many of the chapters in this book, depending heavily on data from case studies in an urban context, reiterate this point. The aim of this chapter is to thus explore the conditions and challenges of improving public transport in rural areas through ICT and AVs. We will also discuss how policy focus needs to change to increase attention to rural areas and give suggestions on concrete policy measures that can be used.

When discussing rural areas in this chapter, we are departing from the degree of urbanisation classification (DEGURBA) distinguishing between cities, towns and suburbs and rural areas. According to this definition, a rural area is defined by a population density that is less than 300 inhabitants per square kilometre (Eurostat, 2018). By ICT, we are referring in a general sense to the possibilities enabled by new technologies for improved information sharing, which can be utilised for matching mobility needs, and making pooling of mobility resources possible. This includes car sharing arrangements, upgraded Demand Responsive Transport (DRT) services and other forms of shared mobility arrangements powered by new ICT (Finger & Audouin, 2018). The definition of AVs in this chapter corresponds to levels 4 and 5 according to the SAE standard (SAE, 2018). Due to reasons elaborated further in the chapter, we argue that the realisation of potential benefits of AVs in rural areas is dependent on the availability of AVs fully capable of operating without a driver.

The chapter is conceptual in its focus and is based on an overview of research and experience on transport in rural areas and on ICT and AVs in public transport. We also draw on the results of two research projects at Lund University, one on the effects of low-carbon transport policies on rural areas (Winslott Hiselius et al., 2020), and the other on trials with on-demand public transport services in various parts of the world (Pettersson, 2019). In the chapter, we depart mainly from the situation in Sweden and take many of our examples from Sweden. However, the outlook of the chapter is general and international and we also refer to examples and figures from other countries.

The Need for Public Transport in Rural Areas

Rural areas have a very strong car dependency compared to urban areas and cities. According to Ridderstedt and Pydokke (2017), 76%–80% of travel to and from work is done by car in rural areas in Sweden, while only 1%–7 % is done by public transport. For cities, there are clear visions on how a transition to low-carbon mobility should be achieved with a mix of modal shift to walking, cycling and public transport, reductions in car use, city planning to reduce travel needs and cleaner fuels in existing cars. Many cities have adopted ambitious policy goals to sharply reduce car traffic and put bans on cars (Business Insider, 2018). Similar visions of low-carbon mobility are lacking for the countryside. In a survey of Swedish municipalities, it was shown that around 90% of the medium to large-sized cities had goals of reducing car traffic, while the number in rural municipalities was around 30% (Hansson, Pettersson, Khan & Hrelja, 2018.) Similarly, all urban municipalities had goals of increasing public transport while this existed in only half of the rural municipalities.

Public transport has emerged as a main solution in strategies for a sustainable transport system. In Sweden, investments in public transport have increased substantially over the past 10 years, and the sector has adopted a goal of doubling the market share of public transport (from 20% to 40%) by 2030. From the 1970s to the mid-2000s, the share of public transport was fairly stable around 20%, but between 2005 and 2015, there was an increase in the share to around 25% (Svensk Kollektivtrafik, 2018). The increase has come mainly in larger cities and in regional routes between urban centres. In the last decade, public transport planning has had a strong focus on efficiency and increasing volumes by investing in strong routes where many people travel. This approach to public transport planning is an internationally well-established best practice (Nielsen et al., 2005).

The role of public transport in rural areas has received much less focus in both research and policy practice (Hansson, Pettersson, Svensson, & Wretstrand,

2019.). As we see it, public transport in rural areas can serve two main purposes. On the one hand, it can serve as an alternative to car travel and thus be a means to lower the environmental impacts of transportation. On the other hand, public transport can be important to secure accessibility for people with limited or no access to a car, making it possible for them to live in rural areas (Winslott Hiselius et al., 2020). While these two goals can be combined, there is also a potential policy conflict between increasing the competitiveness for public transport in strong routes and maintaining good access on less travelled routes (Khan, Hrelja & Pettersson, forthcoming). Lindgren and Berg (2017) also show that public transport in rural areas in general has been deprioritised in Sweden because of tough requirements on cost coverage and a focus on increasing volumes. For rural traffic, there has been a marked priority of strong routes, increased frequency on these routes and straighter shorter routes, in order to increase travel volumes and make public transport more competitive. The downside of this is that less profitable routes, often in smaller towns, are being cancelled, which affects people who are dependent on these routes. In a Swedish context, Ridderstedt and Pydokke (2017) have studied travel patterns in rural areas and found that those who use public transport are mainly young people, with school trips constituting a large share of total trips. The study also found some correlation between income and gender, where lower income groups and women use public transport to a higher degree. About one quarter of the adults who use public transport do not have a driver's licence making them highly dependent on public transport (Ridderstedt & Pydokke, 2017). Consequently, questions about accessibility and equity become central, since some groups risk being excluded from social activities and public services or might not have the possibility to continue living in the countryside (Berg & Thoresson, 2017).

Challenges and Solutions for Public Transport in Rural Areas

While public transport can play a vital role in rural areas, there are also important challenges to maintaining a high service level at a reasonable cost. In the following section, we will describe the main challenges and how they are handled today. Later in the chapter, we will discuss how ICT and AVs could help in addressing them.

Providing public transport in rural areas is imbued with many similar challenges, regardless of national context (International Transport Forum (ITF), 2015a). Challenges include low population densities and increasing competition from cars, which means that there is a limited base of potential passengers. Increasing urbanisation also entails a decreasing and ageing rural population (Berg & Thoresson, 2017). Furthermore, travel demand in rural areas is characterised by being irregular in time and space. It is genuinely difficult to design fixed line public transport networks under such conditions, and the consequence is often a negative feedback loop where infrequent services, not meeting the travel demand, result in few passengers and low-cost coverage. Low ridership, low cost coverage and the subsequent need for high levels of subsidies are in turn often the basis for further service reduction, or ultimately a complete discontinuation of services (ITF, 2015a). Considering this challenge, it should be noted that labour costs for drivers constitute a large part of total costs for bus services (50% according to Jansson, 1980, and 70% according to Currie, 2018).

DRT has often been brought forward as a possible solution to address the need for public transport in rural areas. DRT differ from conventional public transport services by not operating along fixed-route, with fixed-schedules. Instead, DRT services are designed to respond to individualised requests or demands for transportation service (TCRP, 2008). However, some examples from Sweden illustrate that the practical experience with DRT traffic has hitherto been meagre. DRT services, both those open to the general public and special services, constitute a very small part of total travel with local and regional public transport in Sweden. Of the 1.4 billion trips made in 2013, DRT accounted for only 0.6 per mille. In terms of vehicle kilometres, DRT accounted for 2 per mille of a total of 808 million kilometres. The share of costs was 5 per mille of SEK 33 billion (Trafikanalys, 2015).

In Sweden, the average cost of a DRT trip is SEK 182 per trip (Trafikanalys, 2015). This can be compared with the average cost of a trip with the public subsidised public transport in 2015, which was SEK 29 (Trafikanalys, 2016). However, in terms of individual lines, there are large variations; for example, one fixed route bus line in a small rural town in Sweden reported a cost of SEK 245 per trip (Skaraborgs Läns Tidning, 2016). These examples illustrate the problems with providing public transport in rural areas described earlier in this chapter. Regardless of fixed line, or DRT services, low utilisation, large deficits and the need for significant subsidies have so far been a common outcome. Similar examples can be found in many rural areas in Sweden, as well as internationally (cf. Chowdhury & Giacaman, 2015; Davison, Enoch, Ryley, Quddus, & Wang, 2012).

Concerning DRT services, there are a number of previously known barriers to implementation that can serve as an explanation for the modest results in the Swedish examples. These include inadequate technology, poor matching of service provision and geographical coverage, lack of knowledge about potential users, institutional issues, operational problems, lack of information and differences in attitudes and perceptions between stakeholders (Davison et al., 2012; Mulley, Nelson, Teal, Wright, & Daniels, 2012). We will return to the issue of how ICT and AVs potentially could help overcome some of these barriers in the next section of the chapter.

Apart from DRT, other recurring solutions to tackle the challenges of rural mobility are carpooling, ride-sharing, volunteer-driven minibus services and increased coordination between service traffic (e.g. school buses) and regular public transport (Berg & Thoresson, 2017; ITF, 2015a). As with DRT, ride sharing and carpooling are not new phenomena, but, spurred by advances in ICT, there is a renewed interest in developing such services. This development has also led to a blurring of what is considered public transport. In France, encouraging ride-sharing and carpooling is viewed as an important complement to providing public transport, and it is considered as key element in the government's policies for improving alternatives to private car use in rural areas (ITF, 2015a).

A number of critical challenges with ride-sharing schemes and other volunteer-based systems in rural areas exist. These include reaching a critical mass of users to increase chances of matching travel demand of the users. Voluntary approaches are also often inhibited by the availability to involve a feasible number of volunteers that can make sure that the services are extensive enough to provide an attractive service offering. How to handle the transactions of money for providing services (including handling conflicts with infringing on taxi services) is a critical issue. How to integrate these more informal means of public (or shared) transport with the formal public transport system is another challenge (ITF, 2015a). If such barriers could be overcome, it is clear that these kinds of initiatives would be more attractive from the perspective of public authorities since they hold the promise of a low-cost alternative to improving the alternatives to the private car in rural areas. However, based on the development with the protests of the yellow vests, it is safe to assume that the ambitions to support alternatives to private cars in rural France have so far had limited success.

ICT and AVs as a Solution to Public Transport in Rural Areas?

In this section, we will critically discuss in what ways ICT and AVs could provide a solution to the challenges of public transport in rural areas and what the possible limitations are. We discuss the two concepts separately. Although they are connected, they mainly address different challenges: ICT can make it easier to develop more flexible transport solutions, while AVs can primarily lower driver costs, which is a large share of total costs for public bus transport.

ICT and Rural Public Transport

Different forms of ICT are already extensively used in transport systems to improve information sharing and enable new transport services, such as electronic travel planners, including ticketing platforms, and integration with new mobility providers such as bike sharing and electric scooters (see other chapters in this book, e.g. Fearnley, 2020, Chapter 10) In relation to public transport in rural areas, ICT could improve the functioning of DRT systems as well as making other rural mobility schemes more attractive and flexible. A Swedish research project is focussing on launching trials in rural areas in 2020, where a digital platform is developed to allow for planning and booking different travel services, such as car-sharing, public transport, carpooling and cargo bike rental (Västra Götalandsregionen, 2019).

One example of an ICT application in rural public transport is Plustur, a service launched to augment the existing DRT services in Northern Jutland, Denmark. The service is part of the regional transport authority's strategy to improve accessibility in the rural parts of the region, specifically by addressing the first- and last-mile problem. The Plustur service is only available through the public transport journey planner and trips can be ordered from an address to a bus/train stop, or vice versa. As such, the service is available whenever the public transport system is operating. The integration between DRT and the general public transport system through the journey planner is a good example of an ICT enabled improvement (Pettersson, 2019). Plustur results in a marked increase in accessibility to public transport for the rural parts of the region. According to the regional Public Transport Authority (PTA) in Northern Jutland, before the launch of Plustur, 6% of houses (in the entire region) were located within 500 metres of a public transport stop. With the launch of Plustur, the regional PTA argues that the entire region has 100% accessibility to public transport (Nordjyllands Trafikselskab, 2018).

Another example of an ICT-augmented DRT service is Breng flex in the Netherlands, which operates in and around the towns of Arnhem, Nijmegen and Molenhoek. This DRT service was developed to replace low performing fixed bus routes. Like Plustur, the Breng flex service is integrated with the general public transport system through a national journey planner. A main difference is that Breng flex allows for making journeys without including a conventional bus or train journey. As such, it is not considered a first- and last-mile solution, but rather as the main public transport option in the area (Pettersson, 2019).

What can ICT address, and what are other problems that need other measures (e.g. legal, social, cultural and economic)? The examples of Plustur and Breng flex illustrate that so far, the demand is lower than expected, which in turn means that productivity is low and costs are high (Pettersson, 2019). In these cases, ICT has helped overcome barriers concerning information, making booking convenient, and generally improving the interface between users and producers. But the low levels of use indicate that other barriers concerning economics are still relevant. It could also indicate that there is a lack of knowledge about users, and a mismatch between service offerings and the needs for mobility.

AVs and Rural Public Transport

Since fully AVs in public transport services still are under development (as is the case generally for AVs), studies in this field are often modelling possible effects of introducing AVs, or publications that discuss the effects of AVs, on a more conceptual level. Chapter 5 by McLeod, Curtis, and Stone (2020) in this book provides a good overview of different modelling studies. According to the results of some of these studies, the introduction of shared AVs will have a profound impact on urban transport systems, by contributing or counteracting congestion and replacing private-owned cars. According to ITF (2015b), 90% of conventional cars could be made redundant, under 'certain circumstances'. Another positive outlook is provided by Shen, Zhang, and Zhao (2018) who conclude that possible effects of introducing AVs in the Singaporean public transport system include enhanced service quality, congestion reduction, improved financial sustainability of public transport and a more efficient utilisation of bus services. It can, however, be concluded that none of the approximately 50 studies in the overview made by McLeod et al. (2020) have an explicit focus on rural areas as defined in this chapter. Nevertheless, some studies have a geographical scope, for example, a metropolitan region (Boesch, Ciari, & Axhausen, 2016; ITF, 2017), or a whole country (Meyer, Becker, Bösch, & Axhausen, 2017) that include areas that correspond to what we here refer to as rural.

Conclusions from modelling studies addressing effects on rural areas include that combining shared AVs as feeder services with existing high capacity public transport contribute to a radical transformation of access to jobs across rural areas of the Lisbon metropolitan region (ITF, 2017). Similarly, Meyer et al. (2017, p. 80) conclude that the proliferation of AVs could result in 'a quantum leap' in accessibility in Switzerland, and the effects are especially pronounced in rural areas and small cities. A difference compared to ITF (2017) is that Meyer et al. (2017) assume that shared AVs will replace all existing public transport, expect in the major cities.

As discussed earlier in the chapter, key challenges for public transport in rural areas are fragmented demand in time and space, which makes it costly to operate. A key question then is whether AVs could lower the cost of rural public transport. Given that a significant share of the cost for current public transport services is allocated to labour, there is a theoretical potential that AVs could reduce the cost of services in low demand contexts. The potential for cost savings requires highly advanced AVs (corresponding to level 5) which means that there is no need for a driver. This assumption is, in turn, based on that the cost for autonomous technology does not offset the cost savings from not requiring a driver. Since level 5 AVs are yet to be developed, it is not possible to empirically verify that autonomous buses would result in the assumed cost savings, but according to a modelling study by Zhang, Jenelius, and Badia (2019), fully autonomous buses exhibit a significant potential for cost savings, even when assuming very high capital costs for the AVs. Likewise, a study by Abe (2019) modelling the impacts on cost of introducing autonomous buses and taxis in Japanese metropolitan transport systems showed cost reductions of 6%–11% for bus trips, 13%–37% for bus/taxi trips with access to taxi and 44%–61% for taxi trips. It should be pointed out that there are not many publications focussing specifically on AVs in rural areas, and the assessment of costs needs to be critically addressed in further research.

Studies discussing AVs in rural areas on a more conceptual level include two reports by consultancy groups WSP (2016) and Roland Berger (2018). These reports are examples of publications focussing specifically on the potential of AVs in public transport in a rural context. Apart from cost savings, there is also an added dimension about improving mobility options and reduce isolation for segments of the population with limited access to cars, such as the elderly, young people and people with disabilities. It is also argued that new transport services enabled by AVs (e.g. combining passenger transport with logistics services) will have additional benefits, such as opening up opportunities for rural business (WSP, 2016). The potential to attract more visitors to rural tourist destinations, which could improve the economic vitality of rural areas, is also discussed in Roland Berger (2018). It is claimed that rural AVs combining traditional route services with last-mile services could result in commercially viable public transport services by way of cutting labour costs, in combination with increased use and increased fare revenues. In a recent Master's thesis, Norman (2019) assesses the potential for improving public transport in rural areas through the use of driverless vehicles in a Swedish context. Three mobility concepts were identified (shared on demand taxis-service, traffic in a delimited area, and first and last mile with straightened bus routes), and these were analysed using modelling based on real-world case locations. The results show that AVs could contribute to increased flexibility and accessibility of rural populations but that sustainability impacts depend on which travels are substituted (Norman, 2019). The case for AVs also depends heavily on the actual cost savings. While these reports and studies point to possible benefits, there is a need for more research and pilot trials to critically explore and assess both the potential and challenges of AVs in rural public transport.

Roland Berger (2018) raises a number of arguments on why rural settings have advantages compared to urban settings for piloting projects with AVs. One advantage is that testing on roads in the countryside will be easier due to a less complex traffic environment, which will facilitate building consumer confidence in the service. Other advantages include an easier target audience (the elderly being a core market), less demand, and also less demanding passengers in terms of expectations on needing to travel during a specific time of day. As of today, pilot trials with AVs in public bus transport have focussed mainly on urban settings. Within the EU-project CityMobil2 small driverless buses (for 10 passengers) were tested in real settings in seven cities in Europe (Allesandrini, 2018; CORDIS, 2016). Further examples of ongoing trials are in Stockholm (Nobina, 2018), Gothenburg (Chalmers, 2018), Berlin (CBW, 2019) and Singapore (The Strait Times, 2019). There are very few examples of rural projects and, in fact, the only ones we have found in the literature are trials in Japan (Apolitical, 2018) and Australia (Business Insider, 2019), where small driverless shuttle buses serve the transport needs of the elderly in rural areas. These are very limited trials, and there is as of yet no information on outcomes or lessons learned.

When discussing the potential of ICT and AVs to improve mobility for rural areas, a key question is what technology in itself can contribute with, and conversely, what kind of barriers are independent of technology. The account above has provided some examples of currently ongoing ICT enabled attempts at improving DRT and other forms of collective mobility improvements. The examples of ICT augmented DRT services illustrate that while technology can help to alleviate some types of barriers, for example, improved information between users and operators, thus far there is no evidence that new technology makes DRT cheaper to operate. Here, AVs could play an important role based on the critical assumption that it is possible to develop level 5 vehicles without a need for a driver, capable of operating in a rural context. The results from modelling studies seem to indicate that there is a great potential to increase accessibility for rural areas, while at the same time decreasing costs for operating public transport services. It should, however, be noted that there are good reasons for being cautious to the very optimistic scenarios sometimes described as the likely result of AVs. Apart from relying on a level of technological development, which is still in the future, another point is made by Currie (2018) who argues that modelling studies of AVs make biased assumptions about the propensity to share vehicles. According to Currie (2018) there is little (if any) evidence that the hype around 'the sharing economy' has had any impact in the transport sector beyond a superficial level of labelling some new mobility services as 'shared'. Thus, more research and knowledge is needed in order to explore the promises of ICT and AVs for public transport in rural areas.

Policy Suggestions for New Solutions to Public Transport in Rural Areas

First and foremost, we argue that there is a need for a shift in policy focus. The account of public transport in rural areas in this chapter has highlighted that in order to address the existing challenges, ICT and AVs could be important, but given the dominant focus on urban areas, there is a lack of knowledge of concrete measures and policy suggestions for how this potential could be realised. Developing solutions for rural areas will most likely not be of commercial interest. The current focus on urban areas is generally informed by ideas about win-win situations for the public and private sector. The volumes of urban transport represent the business opportunities and therefore attract the interest of investors and developers, but as we have argued above, the social need is stronger in rural areas. Developing mobility solutions based on social needs is different to developing commercially viable services. Therefore, we argue that the state needs to focus more on rural areas in R&D funding, and there is also a need for addressing regulatory and organisational barriers.

On a more concrete level, we argue that since both technologies and services are currently developing, it is important not to focus only on policy instruments pertaining to technology per se. The technologies enabling new services dependent on ICT and AVs are means that could be used to improve rural public transport, but policy instruments making regulatory and organisational changes, as well as pilot schemes, could be more important in the short term to pave the way for improving existing, and establishing new services.

Organisational Changes and Removal of Regulatory Barriers to Integrate Different Services

It is common practice to divide different types of public transport services in terms of who can use them. Systems may, for instance serve the general public, or they may serve only specifically defined groups, such as the elderly, persons with disabilities or school children (TCRP, 2008). Often different organisations have responsibility for different services, which can be one explanation for fragmented, and for users' unattractive systems. According to Westerlund (2016), the 'Danish model' for organising and managing DRT services could serve as a role model for an organisational structure facilitating the integration and coordination of various regional and local passenger services. The 'Danish model' is characterised by a specific national organisation, FlexDanmark, responsible for coordinating a wide range of different DRT services offered by the regional public transport authorities. FlexDanmark is jointly owned by the regional public transport authorities, and this organisational solution facilitates a high level of competence, and economies of scale in the efforts to develop and implement technology. The coordinated development also means a consistency in service offerings, and interfaces with the users, and facilitates integration with fixed route public transport in the Danish regions (Westerlund, 2016).

Combining the transport of passengers and freight, which WSP (2016) suggested as a potential benefit of AVs in rural areas, may also require changes in regulatory frameworks. The requirements for professionally carrying out person and freight transport are typically linked to different regulations and require different forms of permits. In the Swedish context, different levels of value added tax on passenger (6%) and freight transport (25%) also add significant complexity to tax accounting and act as a barrier to integrating passenger and freight transport. An overview of possibilities to harmonise regulatory frameworks and tax levels for passenger and freight services in rural areas could therefore be of importance to facilitate the integration of passenger and freight transport services.

User-centred Design of Public Transport Services in Rural Areas

One well-known barrier to DRT in rural areas is a mismatch between service areas and the mobility demands of people (Mulley et al., 2012). As such, an important policy instrument could be to develop methods and arenas for involving potential users in the design of services. Sarolis (2015) does, for instance, argue that the long-term success of a DRT service is entirely dependent on the involvement of everyone in the local community. Developing methods and areas for facilitating such involvement will probably require that someone, for example, a regional PTA, allocates sufficient resources (staff, knowledge and facilities). However, since there is a lack of knowledge on what methods and arenas deliver the best results, there is also a need for funding research and pilot projects about user-centred design of public transport in rural areas.

An important step in increasing user involvement in designing services could be to utilise data from ride sharing systems, which would provide knowledge about users, for example, regarding spatial and temporal patterns of travel, both by logging executed trips and non-satisfied requests for trips. This knowledge could in turn be used in deliberations of the potential to scale up voluntary, small scale ride sharing services to a more formalised publicly operated DRT system. Promoting interoperability, and sharing of data between different ride sharing platforms, as well as the general public transport system could therefore be important. This suggestion also implies that someone needs to allocate resources to ensure data are handled and analysed in a way that can be useful for developing knowledge about travel demand in specific rural areas. Once again, we argue that public organisations, such as regional public transport authorities, would have a natural role to play here, and that the funding of pilot projects and trials could be important policies.

For rural ride sharing services, key problems include reaching a critical size, and enrolling a sufficient number of volunteers in order to provide a realistic alternative to using the private car (Sarolis, 2015). By default, services operating in a non-profit mode have limited resources and may therefore suffer from being unknown to potential users. Here, a possible policy instrument could be information campaigns about the existence of services and how they work. Another measure to increase the attractiveness of rural ride sharing could be dedicated infrastructure, such as park-and-ride facilities connecting to public transport.

Pilot Schemes to Test AVs for Public Transport in Rural Areas

There are of course significant barriers concerning AV technology, which at present is under development. Vehicles capable of operating safe autonomous public transport services in rural areas, where requirements of relatively high operational speeds are necessary, are yet to be developed. Still, in order to develop AV for public transport in rural areas, there is a need to continuously test vehicles and systems in real-life settings. As we have seen above, there is an increasing number of test schemes for AVs in public transport, but they are almost exclusively focussing on urban contexts. There is a need to expand pilot schemes to other settings such as small towns and rural areas to test the feasibility of AVs and learn about the challenges that arise. Still, this should be done with the awareness that there is a long way to go for fully automated vehicles that can deliver on the promise to reduce operation costs. An initiative launched by the US Department of Transportation in 2019 of awarding a significant research grant ($7.5 million) to study the use of AVs on rural roads is one good example of a state led response to the challenge of AVs in rural areas. Based on the premise that 97% of US roads are classified as rural, while testing has so far largely been confined to urban areas, this project aims to increase the understanding of specific challenges of operating AVs in rural areas, including, for example, poor road network standard (Smartcitiesdive, 2019).

Conclusions

This chapter has argued that there is a need to include rural areas much more in sustainable transport planning since they are more car-dependent than urban areas, and in order to include all parts of society in the transition to low-carbon mobility. Public transport can fulfil an important role in rural areas to reduce car dependence and increase access for different groups, but there are also important challenges to public transport today such as low rates of passengers, high costs and lack of flexibility. In the chapter, we have critically discussed to what extent ICT and AVs can help tackle the challenges of rural public transport and provide more feasible services and public transport systems. We find that there are clear potentials of ICT and AVs, but that there is still a great deal of uncertainty and lack of knowledge that needs to be addressed. Therefore, we argue that there is a need to shift policy focus to look for and explore more actively solutions involving ICT and AVs for rural public transport, and that the public sector has a crucial role for this to happen. Concrete policy measures that could be taken in the short term include removal of regulatory barriers for the use of ICT, developing user-centred public transport services and pilot schemes for AVs in rural areas.

References

Abe, R. (2019). Introducing autonomous buses and taxis: Quantifying the potential benefits in Japanese transportation systems. *Transportation Research Part A: Policy and Practice*, *126*, 94–113. https://doi.org/10.1016/j.tra.2019.06.003

Allesandrini, A. (Ed.). (2018). *Implementing automated road transport systems in urban settings*. Oxfor: Elsevier.

Apolitical. (2018). Forget autonomous cars: Rural Japan is betting on self-driving buses. . *Apolitical*, September 25. Retrieved from https://apolitical.co/solution_article/forget-autonomous-cars-rural-japan-is-betting-on-self-driving-buses/. Accessed on November 29, 2019.

Berg, J., & Thoresson, K. (2017). Mobilitet och transportlösningar på landsbygd. En internationell litteraturstudie [Mobility and transport solutions in rural areas. An international literature study], VTI notat 21-2017. Retrieved from https://www.vti.se/sv/publikationer/publikation/mobilitet-och-transportlosningar-pa-landsbygd_1133341

Boesch, P. M., Ciari, F., & Axhausen, K. W. (2016). *Autonomous vehicle fleet sizes required to serve different levels of demand*. Transport Research Board. Published on 1 January 2016, Research Article. https://doi.org/10.3141/2542-13

Business Insider. (2018). 13 cities are starting to ban cars. *Business Insider*, June 1. Retrieved from https://www.businessinsider.com/cities-going-car-free-ban-2017-8?r=US&IR=T#oslo-norway-will-implement-its-car-ban-by-2019-1. Accessed on September 13, 2019.

Business Insider. (2019). Meet BusBot, the driverless bus being trialled by the NSW government that you hail with your smartphone, *Business Insider*, July 12. Retrieved from https://www.businessinsider.com.au/meet-busbot-the-driverless-bus-being-trialled-by-the-nsw-government-that-you-hail-with-your-smartphone-2019-7. Accessed on November 29, 2019.

CBW. (2019). Berlin trials driverless shuttle. *CBW*, September 3. Retrieved from https://cbwmagazine.com/berlin-trials-driverless-shuttle/. Accessed on November 29, 2019.

Chalmers. (2018). First self-driving bus in operation at Chalmers. *Chalmers*, May 3. Retrieved from https://www.chalmers.se/en/areas-of-advance/Transport/news/Pages/First-self-driving-bus-in-operation-at-Chalmers.aspx. Accessed on November 29, 2019.

Chowdhury, S., & Giacaman, N. (2015). En-route planning of multi-destination public transport trips using smartphones. *Journal of Public Transportation*, *18*(4), 31–45. http://dx.doi.org/10.5038/2375-0901.18.4.3

CORDIS. (2016, November 11). CityMobil2, Final Report Summary. Retrieved from https://cordis.europa.eu/project/rcn/105617/reporting/en. Accessed on November 29, 2019.

Currie, G. (2018). Lies, damned lies, AVs, shared mobility, and urban transit futures. *Journal of Public Transportation*, *21*(1), 19–30. http://doi.org/10.5038/2375-0901.21.1.3

Davison, L., Enoch, M., Ryley, T., Quddus, M., & Wang, C. (2012). Identifying potential market niches for demand responsive transport. *Research in Transportation Business & Management*, *3*, 50–61. https://doi.org/10.1016/j.rtbm.2012.04.007

Eurostat. (2018). Methodological manual on territorial typologies, 2018 edition. Retrieved from https://ec.europa.eu/eurostat/web/products-manuals-and-guidelines/-/KS-GQ-18-008

Fearnley, N. (2020). Micromobility – Regulatory challenges and opportunities. In A. Paulsson & C. H. Sørensen (Eds.), *Shaping smart mobility futures – Governance and policy instruments in times of sustainability transitions* (pp. 169–186). Bingley: Emerald.

Finger M., & Audouin, M. (Eds.). (2018). *The governance of smart transportation systems: Towards new organizational structures for the development of shared, automated, electric and integrated mobility*. Cham: Springer.

Hansson, J., Pettersson, F., Svensson, H., & Wretstrand, A. (2019). Preferences in regional public transport: A literature review. *European Transport Review*, *11*, 38. https://doi.org/10.1186/s12544-019-0374-4

Hansson, P., Pettersson, F., Khan, J., Hrelja, R. (2018). Kommunerna och kollektivtrafiken En enkätundersökning av Sveriges kommuners arbete för att stöjda kollektivtrafiken [Municipalities and public transport, a survey of Swedish municipalities work with improving public transport], K2 WORKING PAPERS 2018:6.

International Transport Forum (ITF). (2015a). International experiences on public transport provision in rural areas. Retrieved from https://www.itf-oecd.org/sites/default/files/docs/15cspa_ruralareas.pdf

International Transport Forum (ITF). (2015b). Urban mobility system upgrade: How shared self-driving cars could change city traffic. Retrieved from http://www.internationaltransportforum.org/Pub/pdf/15CPB_Self-drivingcars.pdf

International Transport Forum (ITF). (2017). Transition to shared mobility – How large cities can deliver inclusive transport services. Retrieved from https://www.itf-oecd.org/sites/default/files/docs/transition-shared-mobility.pdf

Jansson, J. O. (1980). A simple bus line model for optimisation of service frequency and bus size. *Journal of Transport Economics and Policy*, *14*, 53–80.

Khan, J., Hrelja., Pettersson, F. (forthcoming). How to increase travel by public transport? Analysis of planning principles and public transport governance in Swedish regions with the highest growth of travel. Under review in *Transportation Research Part A: Policy and Practice*.

Lindgren, H., & Berg, J. (2017). *Lösningar för kollektivtrafik på landsbygd: det pågående arbetet i Sverige* [*Solutions for public transport in rural areas. The ongoing work in Sweden*], VTI rapport 950. Retrieved from https://www.vti.se/sv/publikationer/publikation/losningar-for-kollektivtrafik-pa-landsbygd_1152580

McLeod, S., Curtis, C., & Stone, J. (2020). Planning urban futures for autonomous and shared vehicles: The role of planning support tools as a policy instrument. In A. Paulsson & C. H. Sørensen (Eds.), *Shaping smart mobility futures – Governance and policy instruments in times of sustainability transitions* (pp. 75–103). Bingley: Emerald.

Meyer, J., Becker, H., Bösch, P. M., & Axhausen, K. W. (2017). Autonomous vehicles: The next jump in accessibilities?. *Research in Transportation Economics*, *62*(2017), 80–91. https://doi.org/10.1016/j.retrec.2017.03.005

Mulley, C., Nelson, J., Teal, R., Wright, S., & Daniels, R. (2012). Barriers to implementing flexible transport services: An international comparison of the experiences in Australia, Europe and USA. *Research in Transportation Business & Management*, *3*, 50–61. https://doi.org/10.1016/j.rtbm.2012.04.001

Nielsen, G., Nelson, J. D., Mulley, C., Tegnér, G., Lind, G., & Lange, T. (2005). Public transport. Planning the networks. HiTrans Best Practice Guide 2. Retrieved from http://civitas.no/assets/hitrans2publictransportplanningthe-networks.pdf

Nobina. (2018, October 25). Första självkörande bussarna i reguljär linjetrafik i Barkarbystaden [First self-driving buses in regular traffic in Barkarbystaden]. Retrieved from https://www.nobina.com/sv/sverige/nyheter/Forsta-sjalvkorande-bussarna-i-linjetrafik/. Accessed on November 29, 2019.

Nordjyllands Trafikselskab. (2018). Integrated mobility in rural areas. Presentation by Ole Schleemann, Nordjyllands Trafikselskab at BEST workshop #3, 8 November, Oslo, Norway.

Norman, J. (2019). *Assessing the potential for improving public transport in rural areas by using driverless vehicles*. Master's thesis, Uppsala University. Retrieved from https://www.utn.uu.se/sts/student/wp-content/uploads/2019/10/1910_Joel_Norman.pdf

Pettersson, F. (2019). *An international review of on-demand public transport services*. K2 Working Paper 2019:5. Retrieved from http://www.k2centrum.se/international-review-experiences-demand-public-transport-services

Ridderstedt, I., & Pydokke, R. (2017). Vilka reser med kollektivtrafik i lands- och glesbygd? Demografi och resmönster [Who travels with public transport in rural areas? Demography and travel patterns], VTI notat 13-2017. Retrieved from https://www.

vti.se/sv/publikationer/publikation/vilka-reser-med-kollektivtrafik-i-lands--och-glesb_1088506

Roland Berger. (2018). *Reconnecting the rural. Autonomus driving as a solution for non-urban mobility*. Roland Berger. Retrieved from https://www.rolandberger.com/de/Publications/Reconnecting-the-rural-Autonomous-driving.html

SAE. (2018). Levels of driving automation. Retrieved from https://www.sae.org/news/press-room/2018/12/sae-international-releases-updated-visual-chart-for-its-%E2%80%9Clevels-of-driving-automation%E2%80%9D-standard-for-self-driving-vehicles,

Shen, Y., Zhang, H., & Zhao, J. (2018). Integrating shared autonomous vehicle in public transportation system: A supply-side simulation of the first-mile service in Singapore. *Transportation Research Part A: Policy and Practice*, *113*(C), 125–136. https://doi.org/10.1016/j.tra.2018.04.004

Skaraborgs Läns Tidning. (2016, januari 28). Drygt en resenär per tur [Around one traveller per trip]. Retrieved from https://www.skaraborgslanstidning.se/article/drygt-en-resenar-per-tur/. Accessed on December 12, 2019.

Smartcitiesdive. (2019). Ohio's $17M project to test AVs on rural roads. Retrieved from https://www.smartcitiesdive.com/news/ohios-17m-project-to-test-avs-on-rural-roads/562747/. Accessed on December 12, 2019.

Svensk Kollektivtrafik. (2018). Årsrapport 2018 Kollektivtrafikbarometern [Annual report Public transport barometer], *Svensk Kollektivtrafik*. Retrieved from https://www.svenskkollektivtrafik.se/globalassets/svenskkollektivtrafik/dokument/aktuellt-och-debatt/publikationer/kollektivtrafikbarometern-arsrapport-2018.pdf

TCRP. (2008). National academies of sciences, engineering, and medicine 2008. *Guidebook for measuring, assessing, and improving performance of demand-response transportation*. Washington, DC: The National Academies Press. https://doi.org/10.17226/23112

The Strait Times. (2019). Singapore takes another step towards driverless buses. *The Strait Times*, 3 October. Retrieved from https://www.straitstimes.com/singapore/transport/spore-takes-another-step-towards-driverless-buses. Accessed on November 29, 2019.

Trafikanalys. (2015). Kartläggning av anropsstyrd kollektivtrafik 2013 [Mapping of demand responsive public transport 2013], PM 2015: 6. Retrieved from https://www.trafa.se/kollektivtrafik/kartlaggning-av-anropsstyrd-kollektivtrafik-2013-6269/

Trafikanalys. (2016). Lokal och regional kollektivtrafik 2015 [Local and regional public transport 2015], Statistik 2016:26. Retrieved from https://www.trafa.se/kollektivtrafik/kollektivtrafik/lokal-och-regional-kollektivtrafik-2015-4779/

Västra Götalandsregionen. (2019). KomIland. Retrieved from https://www.vgregion.se/om-vgr/satsningar-och-samarbeten/hallbart-resande-vast/projekt/komiland/. Accessed on December 12, 2019.

Westerlund, Y. (2016). Den globala utvecklingen av storskalig öppen och integrerad flextrafik [The global development of large scale, open and integrated flex traffic]. *Trafikverket*, Publikationsnummer: 2016:076. Retrieved from https://trafikverket.ineko.se/se/den-globala-utvecklingen-av-storskalig-%C3%B6ppen-och-integrerad-flextrafik

Winslott Hiselius, L., Khan, J., Smidfelt Rosqvist, L., Lund, E., Nilsson, L., & Nilsson, M. (2020). En rättvis omställning av transportsystemet. En analys av de sociala effekterna av styrmedel för minskade klimatutsläpp [A just transition of the transport system. An analysis of the social effects of policies for reduced climate emissions] , Bulletin 318:2020, Transport and road, Lund University.

WSP. (2016). Making better places: Autonomous vehicles and future opportunities, WSP, Parsons Brinckerhoff in association with Farrells. Retrieved from https://www.wsp.com/en-KR/insights/autonomous-vehicles

Zhang, W., Jenelius, E., & Badia, H. (2019). Efficiency of semi-autonomous and fully autonomous bus services in trunk-and-branches networks. *Journal of Advanced Transportation*, *2019*, 7648735. doi:10.1155/2019/7648735

Conclusions

Chapter 12

Governance and Citizen Participation in Shaping Futures of Smart Mobility

Claus Hedegaard Sørensen and Alexander Paulsson

ABSTRACT

In this chapter, the authors will summarise the entire book and look ahead. The aim of this book has been to take the calls for governance of smart mobility one step further by analysing and discussing current and future policy instruments to govern smart mobility. The task has been carried out by discussing the why, how and what of policy instruments. So far, the policy instruments governing smart mobility to a large extent are focussed on understanding this new field of mobility, establishing relations and roles between companies and authorities, and making the field governable. What is lacking in this equation are policy instruments that establish the population as citizens with rights, voices and roles. In order to align the smart mobility transition and the transition towards a sustainable society, the authors consider the development of deliberative citizen participation an important initiative and the authors suggest it as an important field for future research.

Keywords: Governance; citizen participation; smart mobility; democracy; policy instrument; sustainable mobility

Introduction

In the acclaimed book *The Age of Surveillance Capitalism*, Shoshana Zuboff (2019) writes about how digital technology is embedded in social and economic systems and that digital technology has developed into a powerful device to monitor and control behaviour in the latter systems. This has become possible because the digital technology is owned by a handful of corporate tech-giants,

Shaping Smart Mobility Futures:
Governance and Policy Instruments in times of Sustainability Transitions, 205–219

doi:10.1108/978-1-83982-650-420201012

whose idea of doing business and creating value is based on gathering, analysing and selling data. For many years, the traditional auto-manufacturers did not pay much attention to these digital tech-giants and their businesses, as digital technology was primarily confined to developing websites and online services. But when Google started developing its own car and began to connect computers with advanced sensory-equipment to enable autonomous driving, it roused the interest of the traditional auto-manufacturers. BMW and Daimler, to mention two major auto-manufacturers, realised that they were potentially seeing a new competitor entering the market, a competitor with more knowledge about the advanced technological ecosystem needed for making autonomous and connected cars a reality (Hubik & Fasse, 2019).

While autonomous and connected cars, shared mobility services, e-scooters and other forms of new mobility are emerging at a fast pace and potentially redrawing the boundaries of the auto-manufacturing and taxi service markets, the outcome of these developments is currently uncertain. On the one hand, new technologies and services do not in themselves constitute a threat to the global climate goals, but they may trigger and promote practices and investments heading in unsustainable directions. On the other hand, new technologies and mobility providers may contribute to achieving the global climate goals, at least if they are governed in that direction by consciously crafted policies and policy instruments. Given the uncertainty of these developments, several authors have called for a new take on the governance of smart mobility. As we mentioned in the introductory chapter, the ambition of this book has been to take the calls for governance of smart mobility one step further by considering the policy instruments used today, as well as the instruments that may be used in the future to govern smart mobility.

In this book, different perspectives on policy instruments of and for smart mobility are discussed. Yet, all chapters draw upon the definition of policy instruments, either implicitly or explicitly, as 'techniques of governance that, one way or another, involve the utilization of state authority or its conscious limitation' (Howlett, 2005, p. 31). Subsequently, a policy instrument is not understood in this book primarily as a rational tool used by a government to achieve a clearly defined objective (Fischer, Torgerson, Durnová, & Orsini, 2015). Because of this, we have included a range of conceptualisations in this book of what a policy instrument is and consists of, from economics and political science to sociology. Also included in this discussion is the choice of policy instrument, as this often reflects current ideological trends and ideas about what a modern instrument looks like and how such an instrument operates.

Below, we will summarise the conclusions from the three parts of the book. In the first part, we ask why there is a need for policy instruments in relation to smart mobility. In the second part, we ask how policy instruments are chosen and developed. In the third part, we ask what the policy instruments are doing and what smart mobility is doing to them. After this, we look ahead and probe the following future-oriented questions: What can be learned from the results and conclusions presented in this book, and what problems and problematisations have been identified for future research projects? In the final section of this chapter, we discuss the social context of policymaking, the role of citizen participation and the prospects of governing smart mobility in the future. We end this chapter and the book by calling for more reflections on the links between governance, deliberation and citizen participation.

Why is there a Need for Policy Instruments?

The discourse on smart mobility is often very optimistic in suggesting that the digitalisation will contribute to solving a range of problems, for example, transport-related emissions, congestion and traffic safety. This is particularly visible in communication from the industry and public authorities (Bjelvenstam, 2018; Herrmann, Brenner, & Stadler, 2018; Seba, 2014), but research publications also share this optimism (Fagnant & Kockelman, 2015). Coupled with an innovation agenda (Hirschhorn, Paulsson, Sørensen, & Veeneman, 2019; Smith, Sochor, & Sarasini, 2018) and great uncertainty about the impact of the future evolution of smart mobility, public authorities often seem reluctant to interfere with and govern the development of smart mobility (Moscholidou, 2020, Chapter 2 in this volume). However, the general trend in the research literature is a call to actively govern smart mobility (Moscholidou, 2020).

The many calls for governance of smart mobility are generally rooted in the potentially negative consequences of the emerging market for smart mobility (Docherty, Marsden, & Anable, 2018; Reardon & Marsden, 2018). When the potential consequences of smart mobility are scrutinised, as in Pernestål, Engholm, Kristoffersson, and Hammes (2020, Chapter 3 in this volume), the effects to a large extent seem to conflict with political objectives, like global climate goals and the Sustainable Development Goals (SDGs). Pernestål et al. (2020) emphasise that positive effects of automated vehicles may be observed on an individual level and for the single vehicle, while the outcomes on a systemic or societal level can be counterproductive and lead to negative consequences. Thus, they suggest that a likely consequence is increased VKT (vehicle-kilometers travelled) and congestion, as well as increased energy consumption and emissions. Though electrification and other fossil-free propulsion technologies may, at least to some extent, decouple VKT from emissions, these authors stress that the production of new vehicles and batteries has environmental consequences. All in all, Pernestål et al. (2020) suggest, in line with many other authors (Docherty et al., 2018; Pangbourne, Mladenović, Stead, & Milakis, 2019; Schiller, 2016; Trafikanalys, 2019), that the introduction of autonomous vehicles can cause conflicts between different sustainability goals, for example, accessibility and climate, which need to be handled politically.

This is discussed by Moscholidou (2020), who shows that English regions seek to collaborate with the new players, and that this appears to be a key response to the lack of regulation. She further argues that new mobility services are managed by using old policy instruments and emphasises a need for rebalancing of the narrative around smart mobility as well as focussed action on local transport goals and policy instruments. Proactive political actions are needed, she claims, to introduce governance of smart mobility, as we are now witnessing a short window of opportunity (Moscholidou, 2020; Reardon & Marsden, 2018). Of course, both the design and implementation of policy instruments for smart mobility is a challenging task (Pernestål et al., 2020), but not engaging in such efforts is a much too risky strategy and cannot be considered a responsible option (Moscholidou, 2020).

How are Policy Instruments Developed?

The question of how policy instruments are developed and accepted is a very central issue in policy practice and extensively studied in research. There is a growing interest in understanding how policy instruments are selected and gain legitimacy. While rational-choice-oriented economists are generally interested in the cost, efficiency and effectiveness of policy instruments, political scientists and other social scientists are often interested in the legitimacy of policy instruments and whether or not they are accepted by the intended target group. One question that has received a lot of attention in this context is why some policies and policy instruments are accepted, while others are not (e.g. Drews & van den Bergh 2016; Dreyer & Walker, 2013). When the intended target group agrees with the ideas and rationalities behind a new policy, they will generally also accept the introduction of connected instruments. However, the opposite may also be true. For example, when the intended target group disagrees with the underlying ideas and rationalities, there will often be opposition to instruments that seek to change established social practices.

By speaking of a 'target group', we indicate that this group is a social category, framed and constructed partly by the policy itself and partly by the values and relationships already at work within this social category. While much work and effort are sometimes placed on identifying the 'right' target group, as this is presumably required to achieve the desired goals, in other cases this is hugely overlooked. Ingram and Schneider suggest that a number of things are taken into account when policymakers want to match a policy with an intended target group, for example:

> effectiveness, ease of implementation, availability of resources, and importantly, elected legislators desire to align themselves positively with widely held public values of how different sort of people should be treated. (Ingram & Schneider, 2015, p. 259)

The recent developments with *the Yellow Vests* movement in France clearly show how a policy is involved in constructing its own target group. Similar developments are seen in Iran, Sweden, Zimbabwe and Ecuador just to name a few examples (*Dagbladet Information*, 2019). In these cases, the target group, or at least parts of it, voiced their discontent. But such direct opposition against increased fuel taxes, which, as it were, intend to correct undesired, external cost by forcing market actors to internalise them in the price, is only one example. There are also cases where the target group disagrees and explicitly opposes the policy, but still accepts the connected policy instruments. In this case, the literature talks about policy acceptance (Hysing & Isaksson, 2015; Isaksson & Richardsson, 2009; Schade & Schlag, 2003). Whether a certain policy instrument is accepted or not may be explained by the fact that the target group trusts the politicians, the authorities or the political system. The governed target group may also accept a policy instrument if it is the result of a political decision-making process that they consider legitimate. Or they may accept it if other people in the target group accept it. The explanations are manifold (Schmidt, 2013; Wallner, 2008).

Furthermore, the origin of a policy as well as which public authority is the owner of a specific policy instrument may also impact on its acceptance. Aligning policy across different tiers of government is crucial for policy acceptance. But this can be difficult when different tiers of government are representing different competences and make use of different policy instruments for various purposes. In his chapter on multi-level governance and smart mobility, Docherty (2020, Chapter 4 in this volume) discusses what issues are delegated from the central level to the regional and local levels of government, and how to achieve policy alignment in such a tiered environment. Unless policy alignment is in place across different tiers of government, there is a risk that policies and policy instruments will not work as intended. Policy alignment can be realised through 'policy moments', where priorities can be set jointly. Docherty argues that priorities for smart mobility must centre on the key issues, such as pricing and the question of who pays, and whether – and if so how – space and road allocation must be reorganised. Should different policy instruments lead in different directions and thus turn out to be unaligned, for example, because different types of behaviours and investments are promoted by different kinds of policy instruments, then policy acceptance will be difficult. Policy alignment is therefore considered an important component for achieving policy acceptance (Pytlik Zillig, Hutchens, Muhlberger, Gonzalez, & Tomkins, 2018).

Policymaking and the selection of a policy instrument are processes involving and drawing upon best-practices, evidence-based decision-making and decision-support tools. In fact, policymakers often try to gain legitimacy for their ideas by referring to a proper decision-making material (Gudmundsson & Sørensen, 2013). Although the constituencies are rarely interested in these details, the press, interest organisations, NGOs or other politicians may question the decisions if adequate and proper investigations supporting the policy are not available. Within the field of mobility, an important planning support tool has been modelling of expected travel behaviour and this has consequently led to certain political decisions (McLeod, Curtis, Stone, 2020, Chapter 5 in this volume). In their chapter, McLeod et al. (2020) show that recent modelling studies of autonomous and shared vehicles have a limited knowledge base, from which future planning may be supported. Thus, they suggest a number of alternative methods, including pluralistic, discursive and transparent methods for planning.

Apart from trust and legitimacy, fairness and equality also appear as important components for understanding how policy instruments are developed, used and possibly accepted (Givoni, 2014). If a policy instrument creates distributional effects that are considered unfair, this may obstruct its use and acceptance (even if the process itself to produce the policy is considered legitimate). In their chapter on experimental governance and public values, Kronsell and Mukthar-Landgren (2020, Chapter 7 in this volume) discuss this at some length. They have investigated whether experimental governance lives up to public values, understood as values safeguarded by public authorities, including but not limited to impartiality, democracy and equality. According to Kronsell and Mukthar-Landgren (2020), local authorities can ensure that these values are in place by being aware of, and drawing attention to, who is included and excluded in these experiments. Are only techno-optimists included in the experiments as potential users, or are critical citizens involved? There are many indications that experimental governance is primarily a way to facilitate market development and helping companies to establish themselves on this market, while attempts at changing travel habits or mobility patterns are either not addressed or placed in the background. Although the distributional effects of experimental governance are not explicit, the issue of fairness and equality is clearly there, not least in terms of inclusion and exclusion.

The fact that corporations are benefitting from this kind of experimental governance in the short term is obvious. In fact, experimental governance has lately emerged as a way for governments to implement national innovation policies (Hirschhorn et al., 2019; Smith et al., 2018). But at the same time, there are huge differences between the corporations in this context. On one hand, you have the large auto-manufacturers with their lobby machinery, and increasingly also the tech-giants (Zuboff, 2019) On the other hand, you have the small tech start-ups and their venture capitalist supporters. While some companies benefit from this experimental approach to governance, others do not. Different corporations have different possibilities to participate in experiments and the opportunities to shape, and make use of, public places to test their mobility solutions are not distributed equally. Therefore, it is not surprising that Stone, Ashmore, Legacy, and Curtis (2020, Chapter 6 in this volume) describe corporations in Australia demanding a coherent set of policy instruments and rules-packages for the development of smart mobility. Without such policies and policy instruments and regulations, competition in the emerging market could be distorted, perhaps even with oligopolies arising, which could ultimately disadvantage consumers.

What are Policy Instruments Doing and What Do Smart Mobility Do to Them?

The focus in this section is on the impact of policy instruments as well as the way policy instruments are impacted by smart mobility. Smart mobility may challenge certain current instruments and thereby establish changes in current instruments. Reardon (2020, Chapter 8 in this volume) is an example of this line of thought. She illustrates how current instruments of organisation, treasure and authority can be challenged by smart mobility and thus call forth a need to change the instruments. Thus, smart mobility has the potential to aid the integration of the transport system as well as integration with land use.[1] Smart mobility may also challenge the current system of funding and subventions and undermine parking standards (Reardon, 2020). In this way, smart mobility can provide the impetus for change and calibration of policy instruments.

[1]The case mentioned by Pettersson-Löfstedt and Khan (2020, Chapter 11 in this volume) of a model for integration and coordination of various regional and local passenger services is an example.

The term 'governing capacity' is introduced by Wallsten, Sørensen, Paulsson, and Hultén (2020, Chapter 9 in this volume) to analyse how the governing capacity of some policy instruments may be undermined by smart mobility. They suggest that the state's governing capacity may decrease in terms of its nodality and organisation, while particular features of authority and treasure may increase. Against this background, they suggest that adaptations of the portfolio of state policy instruments are necessary. While the methods and conclusions of Reardon and Wallsten et al. differ, the analyses emphasise that smart mobility will have an impact on current policy instruments, which for better or worse will need to be changed, calibrated or configured in new settings or packages. It should be stressed that smart mobility may also make some policy instruments easier to apply, such as congestion charging (Reardon, 2020), or establish new types of instruments.

So, the currently used policy instruments may be impacted by smart mobility, but what is the impact of applied or suggested policy instruments on the governing of smart mobility? The target groups of the policy instruments vary, but they are often first and foremost providers of smart mobility. Fearnley (2020, Chapter 10 in this volume) suggests a long list of policy instruments that seem relevant to regulate e-scooter providers, of which some are already applied in different contexts. Geofencing, zoning, ban, capping on fleet sizes, entry regulations, charges and fees, penalties, compulsory data sharing are on his list, and the target group of these policy instruments would be the e-scooter providers.

For other policy instruments, however, the authorities, being the regulators and service providers are the target group. This is the case when pilot schemes, regulatory sandboxes, evaluations, user involvement in design phases and R&D funding are suggested by Pettersson-Löfstedt and Khan (2020) and Fearnley (2020). These policy instruments establish knowledge and experiences relating to the new mobility services and to proper regulation methods. Fearnley (2020, p. 181) subsequently suggest 'a flexible approach to regulation', and Reardon (2020, p. 150) encourages 'reflexivity and learning'.

A final target group of regulation is the population, where we suggest distinguishing between the population as users and customers on the one hand and as citizens on the other. While users and customers pay for and utilise public and private services, citizens are entitled to rights and participate in dialogue about the public values and services. Many of the current policy instruments that Wallsten et al. (2020) consider valuable in futures of smart mobility aim to impact the behaviour of the users. This is true, for example, for taxes and fees, as well as for regulations of traffic and drivers. Some of the instruments discussed by Fearnley do not only regulate the e-scooter providers but also aim to impact user behaviour, for example, geofencing and zoning systems. In addition, the users are assigned a role as co-designers, when Pettersson-Löfstedt and Khan (2020) suggest developing methods and arenas involving potential users in the design of demand-responsive transport in rural areas. On a more concrete level, they also suggest information campaigns towards potential users as well as park-and-ride facilities to increase the attractiveness of rural ridesharing.

What are the consequences of the policy instruments? What do they do to companies, authorities and the population? To answer this question, we must first acknowledge a difference between policy instruments currently applied to govern smart mobility, and policy instruments that might be applied in the future. At the moment, smart mobility is a new field. Ride-hailing activities like Uber and Lyft have had a significant impact on actual travel behaviour in many cities (Clewlow & Mishra, 2017), and as illustrated by Fearnley (2020), so has the newest sibling, e-scooters. When it comes to ridesharing facilitated by digitalisation[2] and Mobility as a Service we are still mostly experiencing pilot projects with very limited effects on actual travel behaviour (Currie, 2018; WSP, 2019). While autonomous and connected vehicles are discussed and associated with many dreams and nightmares (and the car manufactures are continuously stating that the fully autonomous car is just around the corner), we are only witnessing small pilot tests so far. All in all, smart mobility as a policy field is in its infancy, and authorities at all levels are struggling to find out if and how to regulate. We are therefore seeing many pilots, testbeds, regulatory sandboxes and evaluations.

For some of the providers of smart mobility services, the approach has often been 'begging for forgiveness rather than first asking for permission' (Fearnley, 2020, p. 171). This is an approach that does not acknowledge the role of authorities as rulers, and mobility service providers as the ruled. The approach suggests that the authorities are not willing or able to exercise all the tools at their disposal, in order to govern smart mobility providers. In some cases, the relation is turned upside-down and it is instead the providers who rule. The aim, however, of some policy instruments has been to establish the distinction between rulers and ruled. By banning ride-hailing or e-scooter services, or introducing other restrictions, authorities aim to institute this distinction. However, other types of relations are also established between authorities and companies, for example, when authorities engage in collaboration (Moscholidou, 2020) with ride-hailing, ridesharing and car-sharing companies, as well as with providers of Mobility as a Service. New roles and relations are established while – in many cases – new providers are introduced to the transport sector.

One explanation for a reluctant and cautious approach by authorities is that the authorities in many cases are caught unprepared by the introduction of smart mobility. As shown above, tough regulation and restrictions are introduced in some cases, but collaborative arrangements and soft instruments like pilots and evaluations are probably more common (Moscholidou, 2020). For authorities, a motivation for entering into the smart mobility field is also symbolic, to show that they are modern, aware of new trends and possibilities, and able to handle them (inspired by Røvik, 2011). Linked to this is also the understanding of smart mobility as a tsunami that you cannot escape. For authorities, the policy instruments introduced generate knowledge about the technology, the business models and the stakeholders, and they contribute to learning about how to govern and how to craft future policies. In addition, the policy instruments are positioning the authorities as collaborative partners and regulators of new mobility actors in the market, and in relation to the citizens, the ambition is to be viewed as modern and responsible entities.

[2]We distinguish here between ride-hailing (providing a new trip to the customer) and ridesharing (where no new trips are provided) (Montero, 2018).

In these early stages of the smart mobility transition, the users are probably the least impacted by the policy instruments applied in the field of smart mobility (though they may be highly impacted by the changes that come as a result of the policy instruments, i.e. how smart mobility develops). As illustrated above, some policy instruments are directed towards the users, and a ban of ride-hailing or e-scooters of course also impacts on user behaviour, just as new service options established through Mobility as a Service, ridesharing and ride-hailing do. The users are also invited to fill in surveys contributing to evaluation. But all in all, the populations are not very visible as stakeholders in the smart mobility field. Their roles as users or customers are confirmed in pilot projects, testbeds and evaluations related to the new services, and as citizens they may engage in smart mobility issues at the political level or within the public realm. In a future where smart mobility is developed, providing more services, and the policy field has matured, this could change.

All in all, the policy instruments within the smart mobility field are to a large extent focussed on understanding this new field of mobility, establishing relations and roles between companies and authorities, and making the field governable. What seems almost non-existent are policy instruments that establish the population as citizens with rights, voices and roles in the smart mobility transition. Citizen participation is therefore the topic of the last section.

From Rulers to Ruled: Governance in a New Light

In this final section, we will take stock and look ahead on governance and policy instruments of smart mobility in an era of sustainability transitions. While many of the chapters in this book have started from the point of view of the rulers and asked how current policy instruments are impacted or how new policy instruments can be developed, we now want to turn our attention to the ruled (Zeitlin, 1997).

The ruled is a heterogeneous group that may – or may not – accept a specific policy instrument. Earlier, we discussed the question of when, and under what conditions, an intended target group accepts a policy instrument. Not everyone in the target group must accept the instrument for it to have an impact. If an influential minority of the target group accepts the instrument, then the majority will generally follow (Butera, Falomir-Pichastor, Mugny, & Quiamzade, 2017). In this context, the ruled may include all citizens, car-users, or a specific group of individuals. But the ruled can also include corporations, or a group of corporations, for example, corporations operating in the transport sector, such as smart mobility providers. Irrespective of how the ruled target group is defined at the outset, they are all defined in relation to the policy instruments that are being developed and used by the rulers. The ruled are governed and that is their common denominator.

In democracies, there is a chain of power and accountability between the rulers and the ruled. Ideally, the ruled could shape the policy agenda by electing politicians and so engage in a form of democratic self-rule. Yet, there is some evidence to suggest that the rulers – including career politicians, high-raking officials and policy experts – have merged into an elite, working out of reach of most people and operating autonomously of the chain of power in democracies (Crouch, 2004).

It is contestable to which extent this analysis holds true, but it seems safe to assume that some degree of division between an elite and a ruled target group can be observed. It is in such a context that the development of smart mobility and the making of policy for sustainability transitions must be understood. Without acknowledging this critical context, there is a risk that policy is developed by policy experts at great distance from the ruled target group, and that these policies are not developed taking their point of departure in the lives of the majority of citizens (e.g. Crouch, 2004). There is definitely a need for specialists and expert knowledge in a warming world reliant on highly developed technological systems. Yet, there is often a dividing line among the citizens when it comes to the role of experts in policymaking. On the one hand, there is faith in expertise and scientific results, not least regarding climate change. On the other hand, there are citizens who question the role of science and expert knowledge in policymaking. Strassheim (2014) suggest that we are witnessing a dual movement, where there is the expertisation of democracy and the democratisation of expertise, both happening at the same time.

How, then, could new energy be infused in democratic institutions and so make smart mobility align with policies for sustainability transitions? As the ongoing climate crisis requires radical transformations of current society and of lifestyles, collective decision-making processes are necessary. Establishing a legitimate collective decision-making process is impossible, or at least very difficult, beyond democratic institutions. This raises a number of questions related to the locus and focus of transition, as well as the role of expertise and citizen involvement.

While the proponents of the smart mobility agenda (e.g. Bjelvenstam, 2018, Fagnant & Kockelman, 2015, Seba, 2014) usually emphasise the ability of smart mobility to solve a range of problems, including reduction of greenhouse gas emissions, improved mobility, and limiting congestion, how and where technology will be used is an open-ended question. There is a literature suggesting that the core reasons for experimenting with and introducing smart mobility in many countries is related to national innovation policies. Hirschhorn et al. (2019) and Smith et al. (2018), for example, have shown how the Finnish government seems to encourage MaaS developments because they envisage an 'innovation journey'. Obviously, sustainability and innovation are not mutually exclusive. A plethora of innovative and creative ideas and competences will be necessary for a transition towards a sustainable society. The core challenge for policymaking, however, is to make sustainability the primary objective and relate the discussion of policy instruments to the sustainability goals, rather than to smart mobility per se.

As mentioned above, the climate agenda has largely emerged in tandem with the expert rule typical of post-democracy (Crouch, 2004). As the climate agenda is based in climate science, it may come across as complicated to understand for laymen. The expert character of the climate agenda tends to establish a divide between the ruling elite, influenced by climate science and many other lobby groups, and the ruled populations, which may not accept policies that make car-driving difficult or more expensive (e.g. Strassheim, 2014). This is reflected in the social unrest occurring in a number of countries worldwide because of increased fuel prices justified by the climate agenda. At the same time, social movements have emerged, such as Extinction Rebellion and Fridays for Future, which put the climate agenda on the politicians' tables and seem to have some faith in the political institutions, expecting them to manage the looming climate issue. It is in this context that climate experts are working and needed as an indispensable part of the public debate. But how do we overcome the divide between political and corporate elites and the citizens, or – to frame it differently – between the rulers and the ruled?

Despite the presence of technical solutions and relevant proposals of institutional changes, transition ultimately depends on the citizens. To establish transformative transitions, deliberation and citizen participation are needed to invent new technologies and institutional set-ups, as well as to establish the necessary acceptance and legitimacy of policies that target behavioural changes. Political authorities and market actors cannot establish the transition without the citizens, which even is the case in many dictatorships (Mulvad, Larsen, & Ellersgaard, 2017; Nielsen, 2019).

Innovation has increased lately in deliberation and citizen involvement. On a global scale, there have been many different national and/or local initiatives that bridge the gap between the rulers and the ruled. So-called citizen assemblies aimed to advise the political authorities on climate issues have been introduced in the Spanish capital Madrid, in various Canadian cities, and at the national level in France (*Dagbladet Information*, 2020; Nielsen, 2019). In June 2019, the British Parliament suggested a citizen assembly to advise on the UK target of zero emission of green-house gases by 2050, and in Denmark a political agreement on new climate legislation has included a paragraph on the establishment of a citizen assembly in connection to developing the first climate action plan in accordance with the new legislation (Danish Ministry of Climate, Energy, and Utilities, 2019). There are many other examples. Despite the positive examples, it should of course be acknowledged that citizens are a heterogeneous group, and participation does not necessarily favour the climate agenda. The digital development of society is argued to make citizen participation easier, and the term 'polis-digitocracy' has been introduced (Lawrence, Ventura, Doody, & Peracio, 2019) as an umbrella term for citizen participation facilitated through digitalisation. All in all, these developments suggest a shift in current smart mobility initiatives considering inhabitants in a city or country as most of all consumers or users to rather consider them as citizens.

When it comes to the transport sector, local land use plans in some countries require dialogue with the inhabitants in the area. But in political decisions on national roads or railways, taxation, legislation on vehicles or fuel, transition to smart mobility, etc., citizen participation is not normally a requirement and the experiences of participation within this field are few. The mobile character of the field seems to prevent citizen participation. Against this background, a future challenge for research is to study how citizen commitment can be introduced in the field of mobility through institutional amendments to existing democratic institutions. Digitalisation may for good or bad have a role to play in this, though it may also increase disempowerment as suggested by Zuboff (2019). It is important in our view to analyse and discuss the advantages, disadvantages and concrete possibilities of introducing citizen participation in the field of smart mobility and revitalise democracy.

The aim of this book has been to take the calls for governance of smart mobility one step further by analysing and discussing current and future policy instruments to govern smart mobility. The task has been carried out by discussing the why, how and what of policy instruments. So far, the policy instruments governing smart mobility to a large extent are focussed on understanding this new field of mobility, establishing relations and roles between companies and authorities, and making the field governable. What is lacking in this equation are policy instruments that establish the population as citizens with rights, voices and roles (and not just as users or customers) in the smart mobility transition and the transition towards a sustainable society. Based on this, we consider the development of deliberative citizen participation an important initiative and we suggest it as an important field for future research.

Acknowledgements

Work on this chapter is funded by three related research projects broadly focussing on governance of smart mobility and funded by the Swedish Knowledge Centre for Public Transport (K2), the Swedish Innovation Agency (Vinnova), as well as the Swedish Energy Agency. The chapter is inspired by the vibrant research environment at K2, and not least by discussions at a research seminar for all authors of this volume that took place in Lund, Sweden, in September 2019. Louise Reardon (University of Birmingham) and Jamil Khan (Lund University) have reviewed a draft version of the chapter and provided excellent and challenging feedback. We are grateful for all the support and inspiration we have received.

References

Bjelvenstam, J. (2018). Vägen till självkörande fordon [The road to self-driving vehicles] (SOU 2018:16). Retrieved from https://www.sou.gov.se/?s=2018%3A16&limiter=allt. Accessed on December 10, 2019.

Butera, F., Falomir-Pichastor, J. M., Mugny, G., & Quiamzade, A. (2017). Minority influence. In S. G. Harkins, K. D. Williams, & J. Burger (Eds.), *The oxford handbook of social influence*. Oxford: Oxford University Press. doi:10.1093/oxfordhb/9780199859870.013.11

Clewlow, R. R., & Mishra, G. S. (2017). *Disruptive transportation: The adoption, utilization, and impacts of ride-hailing in the United States*. Research Report – UCD-ITS-RR-17-07, Davis, UC Davis, Institute of Transport Studies. Retrieved from https://escholarship.org/content/qt82w2z91j/qt82w2z91j.pdf?t=pzeswg. Accessed on January 15, 2020.

Crouch, C. (2004). *Post-democracy*. Cambridge: Polity Press.

Currie, G. (2018). Lies, damned lies, AVs, shared mobility, and urban transit futures. *Journal of Public Transportation, 21*(1), 19–30.

Dagbladet Information. (2019). Verden er i oprør. Hvorfor? [The world is rebelling. why?] *Dagbladet Information*, December 27, 1, 6–7, and 10.

Dagbladet Information. (2020). Macrons borgersamling skal finde socialt retfærdige klimaløsninger [Macron's citizen assembly aims to find socially fair climate solutions]. *Dagbladet Information*, January 15, 8-9.

Danish Ministry of Climate, Energy, and Utilities. (2019). Aftale om klimalov af 6. December 2019 [Agreement on climate bill of December 6, 2019]. Retrieved from https://kefm.dk/media/12965/aftale-om-klimalov-af-6-december-2019.pdf. Accessed on January 15, 2020.

Docherty, I. (2020). Crafting effective policy instruments for 'smart mobility': Can multi-level governance deliver?. In A. Paulsson & C. H. Sørensen (Eds.), *Shaping smart mobility futures – governance and policy instruments in times of sustainability transitions* (pp. 57–73). Bingley: Emerald.

Docherty, I., Marsden, G., & Anable, J. (2018). The governance of smart mobility. *Transportation Research Part A: Policy and Practice*, *115*, 114–125.

Drews, S., & van den Bergh, J. C. J. M. (2016). What explains public support for climate policies? A review of empirical and experimental studies. *Climate Policy*, *16*(7), 855–876.

Dreyer, S. J., & Walker, I. (2013). Acceptance and support of the Australian carbon policy. *Social Justice Research*, *26*(3), 343–362.

Fagnant, D. J., & Kockelman, K. M. (2015). Preparing a nation for autonomous vehicles: Opportunities, barriers and policy recommendations. *Transportation Research Part A: Policy and Practice*, 77, 167–181.

Fearnley, N. (2020). Micromobility – Regulatory challenges and opportunities. In A. Paulsson & C. H. Sørensen (Eds.), *Shaping smart mobility futures – Governance and policy instruments in times of sustainability transitions* (pp. 169–186). Bingley: Emerald.

Fischer, F. Torgerson, D., Durnová, A., & Orsini, M. (2015). Introduction to critical policy analysis. In F. Fischer, D. Torgerson, A. Durnová, & M. Orsini (Eds.), *Handbook of critical policy studies* (pp. 1–24). Cheltenhamn: Edward Elgar Publishing.

Givoni, M. (2014). Addressing transport policy challenges through policy-packaging. *Transportation Research Part A: Policy and Practice*, *60*, 1–8.

Gudmundsson, H., & Sørensen, C. H. (2013). Some use—Little influence? On the roles of indicators in European sustainable transport policy. *Ecological Indicators*, *35*, 43–51.

Herrmann, A., Brenner, W. M, & Stadler, R. (2018). *Autonomous driving. How the driverless revolution will change the world*. Bingley: Emerald Publishing Limited.

Hirschhorn, F., Paulsson, A., Sørensen, C. H., & Veeneman, W. (2019). Public transport regimes and mobility as a service: Governance approaches in Amsterdam, Birmingham, and Helsinki. *Transportation Research Part A: Policy and Practice*, *130*, 178–191.

Howlett, M. (2005). What is a policy tool? Policy tools, policy mixes and policy implementation styles. In P. Eliadis, M. Hill, & M. Howlett (Eds.), *Designing Government: From instruments to governance* (pp. 31–50). London and Montreal: McGill-Queen's University Press.

Hubik, F., & Fasse, M. (2019). BMW and Daimler put aside rivalry to take on Google and Uber. *Handelsblatt*, February 25. Retrieved from https://www.handelsblatt.com/today/companies/historic-partnership-bmw-and-daimler-put-aside-rivalry-to-take-on-google-and-uber-/24035266.html?ticket=ST-2591454-SbqVqTfpaoAfKdGfO-evg-ap5. Accessed on January 30, 2020.

Hysing, E., & Isaksson, K. (2015). Building acceptance for congestion charges – the Swedish in Stockholm experiences compared. *Journal of Transport Geography*, *49*, 452–460.

Ingram, H., and Schneider, A. L. (2015). Making distinctions: The social construction of target populations. In F. Fischer, D. Torgerson, A. Durnová, & M. Orsini (Eds.), *Handbook of critical policy studies* (pp. 259–273). Cheltenhamn: Edward Elgar Publishing.

Isaksson, K., & Richardsson, T. (2009). Building legitimacy for risky policies: The cost of avoiding conflict. *Transportation Research Part A*, *43*, 251–257.

Kronsell, A., & Mukhtar-Landgren, D. (2020). Experimental governance of smart mobility. Some normative implications. In A. Paulsson & C. H. Sørensen (Eds.), *Shaping smart mobility futures – Governance and policy instruments in times of sustainability transitions* (pp. 119–135). Bingley: Emerald.

Lawrence, S., Ventura, J. L., Doody, L., & Peracio, P. (2019). Polisdigitocracy: Citizen engagement for climate action through digital technologies. *Field Actions Science Reports*, *16*, 58–65. http://journals.openedition.org/factsreports/4353

McLeod, S., Curtis, C., & Stone, J. (2020). Planning urban futures for autonomous and shared vehicles: The role of planning support tools as a policy instrument. In A. Paulsson & C. H. Sørensen (Eds.), *Shaping smart mobility futures – Governance and policy instruments in times of sustainability transitions* (pp. 75–103). Bingley: Emerald.

Montero, J. J. (2018). Regulating transport platforms: The case of carpooling in Europe. In M. Finger & M. Audouin (Eds.), *The governance of smart transportation systems. Towards organisational structures for the development of shared, automated, electric and integrated mobility* (pp. 13–35). Berlin: Springer.

Moscholidou, I. (2020). Steering smart mobility services: Governance and accountability challenges for English local authorities. In A. Paulsson & C. H. Sørensen (Eds.), *Shaping smart mobility futures – Governance and policy instruments in times of sustainability transitions* (pp. 19–35). Bingley: Emerald.

Mulvad, A. M., Larsen, A. G., & Ellersgaard, C. H. (2017). *Tæm eliten – fra magtelite til borgerdemokrati* [Curb the elite - from power elite to citizen democracy]. Copenhagen: Informations Forlag.

Nielsen, J. S. (2019). *Som gjaldt det livet. Til den nye klimabevægelse* [As if as it was for life. To the new climate movement]. Copenhagen: Informations Forlag.

Pangbourne, K., Mladenović, M. N., Stead, D., & Milakis, D. (2019). Questioning mobility as a service: Unanticipated implications for society and governance, *Transportation Research Part A: Policy and Practice*, *131*, 35–49. https://doi.org/10.1016/j.tra.2019.09.033

Pernestål, A., Engholm, A., Kristoffersson, I., & Hammes, J. J. (2020). The impacts of automated vehicles on the transport system and how to create policies that target sustainable development goals. In A. Paulsson & C. H. Sørensen (Eds.), *Shaping smart mobility futures – Governance and policy instruments in times of sustainability transitions* (pp. 37–53). Bingley: Emerald.

Pettersson, F., & Khan, J. (2020). Smart public transport in rural areas: Prospects, challenges and policy needs. In A. Paulsson & C. H. Sørensen (Eds.), *Shaping smart mobility futures – Governance and policy instruments in times of sustainability transitions* (pp. 187–201). Bingley: Emerald.

Pytlik Zillig, L. M., Hutchens, M. J., Muhlberger, P., Gonzalez, F. J., & Tomkins, A. J. (2018). Policy acceptance. In *Deliberative public engagement with science*. Cham: Springer.

Reardon, L. (2020). Smart mobility as a catalyst for policy change towards low carbon mobility? In A. Paulsson & C. H. Sørensen (Eds.), *Shaping smart mobility futures – Governance and policy instruments in times of sustainability transitions* (pp. 139–151). Bingley: Emerald.

Reardon, L., & Marsden, G. (2018). Conclusions: A window of opportunity. In G. Marsden & L. Reardon (Eds.), *Governance of the smart mobility transition* (pp. 155–165). Bingley: Emerald.

Røvik. K.-A. (2011). From fashion to virus: An alternative theory of organizations' handling of management ideas. *Organization Studies*, *32*(5), 631–653.

Schade, J., & Schlag, B. (2003). Acceptability of urban transport pricing strategies. *Transportation Research Part F: Traffic Psychology and Behaviour*, *6*(1), 45–61. https://doi.org/10.1016/S1369-8478(02)00046-3

Schiller, P. L. (2016). Automated and connected vehicles: High tech hope or hype. *World Transport Policy and Practice*, *22*(3), 28–44.

Schmidt, V. A. (2013). Democracy and legitimacy in the European union revisited. *Input, Output and 'Throughput'*, *61*, 2–22.

Seba, T. (2014). *Clean disruption of energy and transportation. How silicon valley will make oil, nucelar, natural gas, coal, electric utilities and conventional gas obsolete by 2030*. Silicon Valley: Clean Planet Ventures.

Smith, G., Sochor, J., & Sarasini, S. (2018). Mobility as a service: Comparing developments in Sweden and Finland. *Research in Transportation Business and Management, 27*, 36–45.

Stone, J., Ashmore, D., Legacy, C., & Curtis, C. (2020). Challenges for government as facilitator and umpire of innovation in urban transport: The view from Australia. In A. Paulsson & C. H. Sørensen (Eds.), *Shaping smart mobility futures – Governance and policy instruments in times of sustainability transitions* (pp. 105–118). Bingley: Emerald.

Strassheim, H. (2014). Politics and policy expertise: Towards a political epistemology. In F. Fischer, D. Torgerson, A. Durnová, & M. Orsini (Eds.), *Handbook of critical policy studies* (pp. 319–340). Cheltenham: Edward Elgar Publishing.

Trafikanalys. (2019). *Uppkopplade, samverkande och automatiserade fordon, farkoster och system – ett kunskapsunderlag* [Connected, collaborative and automated vehicles, crafts, and systems - a knowledge base]. Stockholm: Trafikanalyse. Retrieved from https://www.trafa.se/globalassets/rapporter/2019/rapport-2019_8-uppkopplade-samverkande-och-automatiserade-fordon-farkoster-och-system---ett-kunskapsunderlag.pdf. Accessed on January15, 2020.

Wallner, J. (2008). Legitimacy and public policy: Seeing beyond effectiveness, efficiency, and performance. *Policy Studies Journal, 36*(3), 421–443.

Wallsten, A., Sørensen, C. H., Paulsson, A., & Hultén, J. (2020). Is governing capacity undermined? Policy instruments in smart mobility futures. In A. Paulsson & C. H. Sørensen (Eds.), *Shaping smart mobility futures – Governance and policy instruments in times of sustainability transitions* (pp. 153–168). Bingley: Emerald.

WSP. (2019). *Delad mobilitet idag och i framtiden* [Shared mobility today and in the future]. Stockholm: WSP Sverige. Retrieved from https://www.svenskkollektivtrafik.se/globalassets/svenskkollektivtrafik/dokument/aktuellt-och-debatt/publikationer/rapport---delad-mobilitet-idag-och-i-framtiden-2019.pdf. Accessed on January 15, 2020.

Zeitlin, I. M. (1997). *Rulers and ruled. An introduction to classical political theory from plato to federalists*. Toronto: University of Toronto Press.

Zuboff, S. (2019). *The age of surveillance capitalism – The fight for a human future at the new frontier of power*. New York, NY: Public Affairs.

Index

Note: Page numbers followed by "*n*" indicate footnotes.